Encyclopaedia

of

NUCLEAR SCIENCE, TECHNOLOGY AND ENGINEERING

Encyclopaedia of Nuclear Science, Technology and Engineering

Prof. Lalit Sampat

ANMOL PUBLICATIONS PVT. LTD.
NEW DELHI-110 002 (INDIA)

ANMOL PUBLICATIONS PVT. LTD.

Regd. Office: 4360/4, Ansari Road, Daryaganj,
New Delhi-110 002 (India)
Ph.: 23278000, 23261597

Branch Office: No. 1015, Ist Main Road, BSK IIIrd Stage
IIIrd Phase, IIIrd Block,
Bangalore-560 085 (India)
Tel.: 080-41723429
Visit us at: www.anmolpublications.com

Encyclopaedia of Nuclear Science, Technology and Engineering

First Edition, 2009

ISBN 978-81-261-4128-9

PRINTED IN INDIA

Printed at Balaji Offset, Delhi.

Contents

Preface

Nuclear Science is a nuclear technology designed to extract usable energy from atomic nuclei via controlled nuclear reactions. The most common method today is through nuclear fission, though other methods include nuclear fusion and radioactive decay. All utility-scale reactors heat water to produce steam, which is then converted into mechanical work for the purpose of generating electricity or propulsion.

In 2007, 14% of the world's electricity came from nuclear power. More than 150 nuclear-powered naval vessels have been built, and a few radioisotope rockets have been produced. The concepts and Techniques of Nuclear Science are not elementary in the sense that they are easily mastered. They have evolved through a great deal of experimental and theoretical research and cannot be expressed or explained in simple non-technical terms.

Encyclopaedia of Nuclear Science is an effort to provide an account of Nuclear Science not only suitable as an introduction to this field but also provide a sequential compendium of all the latest developments in the field, whereas the selection of subject matter for the present set has been chiefly according to the syllabi prescribed by the Universities for the students. This set should therefore be useful in an advanced undergraduate course in Nuclear Science, Nuclear Engineering or to anyone else with the indicated preparation who might be interested in Radioactivity.

Author

Chapter 1

Nuclear Science

Nuclear Science is a subfield of chemistry dealing with radioactivity, nuclear processes and nuclear properties. It is the chemistry of radioactive elements such as the actinides, radium and radon together with the chemistry associated with equipment (such as nuclear reactors) which are designed to perform nuclear processes.

This includes the corrosion of surfaces and the behaviour under conditions of both normal and abnormal operation (such as during an accident). An important area is the behaviour of objects and materials after being placed into a waste store or otherwise disposed of.

The study of the chemical effects resulting from the absorption of radiation within living animals, plants, and other materials. The radiation chemistry controls much of radiation biology as radiation has an effect on living things at the molecular scale, to explain it another way the radiation alters the biochemicals within an organism, the alteration of the biomolecules then changes the chemistry which occurs within the organism, this change in biochemistry then can lead to a biological outcome.

As a result nuclear chemistry greatly assists the understanding of medical treatments (such as cancer radiotherapy) and has enabled these treatments to improve. The study of the production and use of radioactive sources for a range of processes.

These include radiotherapy in medical applications; the use of radioactive tracers within industry, science and the environment; and the use of radiation to modify materials such as polymers. The study and use of nuclear processes in *non-radioactive* areas of human activity.

For instance, nuclear magnetic resonance (NMR) spectroscopy is commonly used in synthetic organic chemistry and physical chemistry and for structural analysis in macromolecular chemistry.

MAIN AREAS

Radiochemistry is the chemistry of radioactive materials, where radioactive isotopes of elements are used to study the properties and chemical reactions of non-radioactive isotopes (often within radiochemistry the absence of radioactivity leads to a substance being described as being *inactive* as the isotopes are *stable*).

Radiation Chemistry

Radiation chemistry is the study of the chemical effects of radiation on matter; this is very different to radiochemistry as no radioactivity needs to be present in the material which is being chemically changed by the radiation. An example is the conversion of water into hydrogen gas and hydrogen peroxide.

Study of Nuclear Reactions

A combination of radiochemistry and radiation chemistry is used to study nuclear reactions such as fission and fusion. Some early evidence for nuclear fission was the formation of a shortlived radioisotope of barium which was isolated from neutron irradiated uranium (^{139}Ba, with a half-life of 83 minutes and ^{140}Ba, with a half-life of 12.8 days, are major fission products of uranium).

At the time, it was thought that this was a new radium isotope, as it was then standard radiochemical practice to use a barium sulphate carrier precipitate to assist in the isolation of radium.

More recently, a combination of radiochemical methods and nuclear physics has been used to try to make new 'superheavy' elements; it is thought that islands of relative stability exist where the nuclides have half-lives of years, thus enabling weighable amounts of the new elements to be isolated.

THE NUCLEAR FUEL CYCLE

The chemistry associated with any part of the nuclear fuel cycle, including nuclear reprocessing. The fuel cycle includes all the operations involved in producing fuel, from mining, ore processing and enrichment to fuel production (*Front end of the cycle*).

It also includes the 'in-pile' behaviour (use of the fuel in a reactor) before the *back end* of the cycle. The *back end* includes the management of the used nuclear fuel in either a cooling pond or dry storage, before it is disposed of into an underground waste store or reprocessed.

Normal and Abnormal Conditions

The nuclear chemistry associated with the nuclear fuel cycle can be divided into two main areas, one area is concerned with operation under the intended conditions while the other area is concerned with maloperation conditions where some alteration from the normal operating conditions has occurred or (*more rarely*) an accident is occurring.

Reprocessing

Law

In the USA it is normal to use fuel once in a power reactor before placing it in a waste store. The long term plan is currently to place the civil used power reactor fuel in a deep store. This policy of not reprocessing was started in March 1977 for nuclear weapons proliferation reasons.

The President Jimmy Carter issued a Presidential directive which indefinitely suspended the commercial reprocessing and recycling of plutonium in the USA. This Presidential directive is likely to have been an attempt by the USA to lead other countries by example, but many other nations continue to reprocess spent nuclear fuels.

It is noteworthy that the government under Putin (President of Russia) repealed a law which had banned the import of used nuclear fuel into Russia, this change in Russian law now permits the Russians to offer a reprocessing service for clients outside Russia (In a similar way to that offered by BNFL).

PUREX Chemistry

The current method of choice is to use the PUREX liquid-liquid extraction process which uses a tributyl phosphate/ hydrocarbon mixture to extract both uranium and plutonium from nitric acid. This extraction is of the nitrate salts and is classed as being of a solvation mechanism. For example the extraction of plutonium by an extraction agent (S) in a nitrate medium occurs by the following reaction.

$$Pu^{4+}_{aq} + 4NO_3^-{}_{aq} + 2S_{organic} \rightarrow [Pu(NO_3)_4S_2]_{organic}$$

A complex is formed between the metal cation, the nitrates and the tributyl phosphate, and a model compound of a dioxouranium(VI) complex with two nitrates and two triethyl phosphates has been characterised by X-ray crystallography. When the nitric acid concentration is high the extraction into the organic phase is favoured, and when the nitric acid concentration is low the extraction is reversed (the organic phase is *stripped* of the metal).

It is normal to dissolve the used fuel in nitric acid, after the removal of the insoluble matter the uranium and plutonium are extracted from the highly active liquor. It is normal to then back extract the loaded organic phase to create a *medium active* liquor which contains mostly uranium and plutonium with only small traces of fission products.

This medium active aqueous mixture is then extracted again by tributyl phosphate/hydrocarbon to form a new organic phase, the metal bearing organic phase is then stripped of the metals to form an aqueous mixture of only uranium and plutonium.

The two stages of extraction are used to improve the purity of the actinide product, the organic phase used for the first extraction will suffer a far greater dose of radiation. The radiation can degrade the tributyl phosphate into dibutyl hydrogen phosphate. The dibutyl hydrogen phosphate can act as an extraction agent for both the actinides and other metals such as ruthenium. The dibutyl hydrogen phosphate can make the system behave in a more complex manner as it tends to extract metals by an ion exchange mechanism (extraction favoured by low acid concentration), to reduce the effect of the dibutyl hydrogen phosphate it is common for the used organic phase to be washed with sodium carbonate solution to remove the acidic degradation products of the tributyl phosphate.

New Methods Being Considered for Future Use

The Purex process can be modified to make a UREX (URanium EXtraction) process which could be used to save space inside high level nuclear waste disposal sites, such as Yucca Mountain, by removing the uranium which makes up the vast majority of the mass and volume of used fuel and recycling it as reprocessed uranium. The Urex process is a Purex process which has been modified to prevent the plutonium being extracted.

This can be done by adding a plutonium reductant before the first metal extraction step. In the UREX process, ~99.9 per cent of the Uranium and >95 per cent of Technetium are separated from each other and the other fission products and actinides. The key is the addition of acetohydroxamic acid (AHA) to the extraction and scrub sections of the process. The addition of AHA greatly diminishes the extractability of Plutonium and Neptunium, providing greater proliferation resistance than with the plutonium extraction stage of the PUREX process.

Adding a second extraction agent, octyl(phenyl)-N, N-dibutyl carbamoylmethyl phosphine oxide(CMPO) in combination with tributylphosphate, (TBP), the PUREX process can be turned into the TRUEX (TRansUranic EXtraction) process this is a process which was invented in the USA by Argonne National Laboratory, and is designed to remove the transuranic metals (Am/Cm) from waste.

The idea is that by lowering the alpha activity of the waste, the majority of the waste can then be disposed of with greater ease. In common with PUREX this process operates by a solvation mechanism. As an alternative to TRUEX, an extraction process using a malondiamide has been devised. The DIAMEX (DIAMideEXtraction) process has the advantage of avoiding the formation of organic waste which contains elements other than Carbon, Hydrogen, Nitrogen, and Oxygen.

Such an organic waste can be burned without the formation of acidic gases which could contribute to acid rain. The DIAMEX process is being worked on in Europe by the French CEA. The process is sufficiently mature that an industrial plant could be constructed with the existing knowledge of the process.

In common with PUREX this process operates by a solvation mechanism. Selective ActiNide EXtraction. As part of the management of minor actinides it has been proposed that the lanthanides and trivalent minor actinides should be removed from the PUREX raffinate by a process such as DIAMEX or TRUEX. In order to allow the actinides such as americium to be either reused in industrial sources or used as fuel the lanthanides must be removed.

The lanthanides has large neutron cross sections and hence they would poison a neutron driven nuclear reaction. To date the extraction system for the SANEX process has not been defined, but currently several different research groups are working towards a process. For instance the French CEA is working on a bis-triaiznyl pyridine (BTP) based process.

Other systems such as the dithiophosphinic acids are being worked on by some other workers. This is the *Universal* Extraction process which was developed in Russia and the Czech Republic, it is a process designed to remove all of the most troublesome (Sr, Cs and minor actinides) radioisotopes from the raffinates left after the extraction of uranium and plutonium from used nuclear fuel.

The chemistry is based upon the interaction of caesium and strontium with poly ethylene oxide (poly ethylene glycol) and a cobalt carborane anion (known as chlorinated cobalt dicarbollide). The actinides are extracted by CMPO, and the diluent is a polar aromatic such as nitrobenzene. Other dilents such as *meta*-nitrobenzotrifluoride and phenyl trifluoromethyl sulfone have been suggested as well.

Absorption of Fission Products on Surfaces

Another important area of nuclear chemistry is the study of how fission products interact with surfaces; this is thought to control the rate of release and migration of fission products both from waste containers under normal conditions and from power reactors under accident conditions. It is interesting to note that, like chromate and molybdate, the $^{99}TcO_4$ anion can react with steel surfaces to form a corrosion resistant layer. In this way, these metaloxo anions act as anodic corrosion inhibitors.

The formation of $^{99}TcO_2$ on steel surfaces is one effect which will retard the release of ^{99}Tc from nuclear waste drums and nuclear equipment which has been lost before decontamination (eg submarine reactors lost at sea). This $^{99}TcO_2$ layer renders the steel surface passive, inhibiting the anodic corrosion reaction. The radioactive nature of technetium makes this corrosion protection impractical in almost all situations.

It has also been shown that $^{99}TcO_4$ anions react to form a layer on the surface of activated carbon (charcoal) or aluminium. A short review of the biochemical properties of a series of key long lived radioisotopes can be read on line. It is important to note that ^{99}Tc in nuclear waste may exist in chemical forms other than the $^{99}TcO_4$ anion, these other forms have different chemical properties. Similarly, the release of iodine-131 in a serious power reactor accident could be retarded by absorption on metal surfaces within the nuclear plant.

SPINOUT AREAS

Some methods first developed within nuclear chemistry and physics have become so widely used within chemistry and other physical sciences that they may be best thought of as separate from *normal* nuclear chemistry. For example, the isotope effect is used so extensively to investigate chemical mechanisms and the use of cosmogenic isotopes and long-lived unstable isotopes in geology that it is best to consider much of isotopic chemistry as separate from nuclear chemistry.

Kinetics (Use Within Mechanistic Chemistry)

The mechanisms of chemical reactions can be investigated by observing how the kinetics of a reaction are changed by making an isotopic modification of a substrate. This is now a standard method in organic chemistry. Briefly, replacing normal hydrogens (protons) by deuterium within a chemical compound causes the rate of molecular vibration (C-H, N-H and O-H bonds show this) to decrease.

This then can lead to a decrease in the reaction rate if the rate-determining step involves breaking a bond between hydrogen and another atom. Thus, if the reaction changes in rate when protons are replaced by deuteriums, it is reasonable to assume that the breaking of the bond to hydrogen is part of the step which determines the rate.

Uses Within Geology, Biology and Forensic Science

Cosmogenic isotopes are formed by the interaction of cosmic rays with the nucleus of an atom. These can be used for dating purposes and for use as natural tracers.

In addition, by careful measurement of some ratios of stable isotopes it is possible to obtain new insights into the origin of bullets, ages of ice samples, ages of rocks, and the diet of a person can be identified from a hair or other tissue sample.

Biology

Within living things, isotopic labels (both radioactive and nonradioactive) can be used to probe how the complex web of reactions which makes up the metabolism of an organism converts one substance to another. For instance a green plant uses light energy to convert water and carbon dioxide into glucose by photosynthesis. If the oxygen in the water is labeled, then the label appears in the oxygen gas formed by the plant and not in the glucose formed in the chloroplasts within the plant cells. For biochemical and physiological experiments and medical methods, a number of specific isotopes have important applications.

- Stable isotopes have the advantage of not delivering a radiation dose to the system being studied; however, a significant excess of them in the organ or organism might still interfere with its functionality, and the availability of sufficient amounts for whole-animal studies is limited for many isotopes. Measurement is also difficult, and usually requires mass spectroscopy to determine how much of the isotope is present in particular compounds, and there is no means of localizing measurements within the cell.
- H-2 (deuterium), the stable isotope of hydrogen, is a stable tracer, the concentration of which can be measured by mass spectroscopy or NMR. It is incorporated into all cellular structures. Specific deuterated compound can also be produced.
- N-15, the stable isotope of nitrogen, has also been used. It is incorporated mainly into proteins.
- Radioactive isotopes have the advantages of being detectable in very low quantities, in being easily measured by scintillation counting or other radiochemical methods, and in being localizable to particular regions of a cell, and quantifiable by autoradiography. Many compounds with the radioactive atoms in specific positions can be prepared, and are widely available commercially. In high quantities they require precautions to guard the workers from the effects of radiation—and they can easily contaminate laboratory glassware and other equipment. For some isotopes the half-life is so short that preparation and measurement is difficult.

By organic synthesis it is possible to create a complex molecule with a radioactive label that can be confined to a small area of the molecule. For short-lived isotopes such as ^{11}C, very rapid synthetic methods have been developed to permit the rapid addition of the radioactive isotope to the molecule. For instance a palladium catalysed carbonylation reaction in a microfluidic device has been used to rapidly form amides and it might be possible to use this method to form radioactive imaging agents for PET imaging.

- ^{3}H, Tritium, the radioisotope of hydrogen, it available at very high specific activities, and compounds with this isotope in particular positions are easily prepared by standard chemical reactions such as hydrogenation of unsaturated precursors. The isotope emits very soft beta radiation, and can be detected by scintillation counting.
- ^{11}C, Carbon-11 can be made using a cyclotron, boron in the form of boric oxide is reacted with protons in a (p,n) reaction. An alternative route is to react ^{10}B with deuterons. By rapid organic synthesis, the ^{11}C compound formed in the cyclotron is converted into the imaging agent which is then used for PET.
- ^{14}C, Carbon-14 can be made (as above), and it is possible to convert the target material into simple inorganic and organic compounds. In most organic synthesis work it is normal to try to create a product out of two approximately equal sized fragments and to use a convergent route, but when a radioactive label is added, it is normal to try to add the label late in the synthesis in the form of a very small fragment to the molecule to enable the radioactivity to be localised in a single group. Late addition of the label also reduces the number of synthetic stages where radioactive material is used.
- ^{18}F, fluorine-18 can be made by the reaction of neon with deuterons, ^{20}Ne reacts in a (d,^{4}He) reaction. It is normal to use neon gas with a trace of stable fluorine ($^{19}F_2$). The $^{19}F_2$ acts as a carrier which increases the yield of radioactivity from the cyclotron target by reducing the amount of radioactivity lost

by absorption on surfaces. However, this reduction in loss is at the cost of the specific activity of the final product.

Nuclear Magnetic Resonance (NMR)

NMR spectroscopy uses the net spin of nuclei in a substances upon energy absorption to identify molecules. This has now become a standard spectroscopic tool within synthetic chemistry. One major use of NMR is to determine the bond connectivity within an organic molecule.

NMR imaging also uses the net spin of nuclei (commonly protons) for imaging. This is widely used for diagnostic purposes in medicine, and can provide detailed images of the inside of a person without inflicting any radiation upon them. In a medical setting, NMR is often known simply as "magnetic resonance" imaging, as the word 'nuclear' has negative connotations for many people.

NUCLEAR CHEMISTRY

Nuclear Chemistry deals with the nuclei of atoms breaking apart. Atoms are continually undergoing decay. When studying nuclear chemistry, there is a typical format used to represent specific isotopes.

↓ Mass number

$^{238}_{29}U$ ← Element

↑ Charge or atomic number

Nuclear equations are typically written in the format shown below. There are 5 different types of radioactive decay.

- Alpha decay follows the form:

$$^{X}_{Z}A \rightarrow {}^{4}_{2}He + {}^{X-4}_{Z-2}B$$

Alpha particle ↲

Where A is the parent isotope (the atom being broken apart) B is the daughter isotope or the isotope formed. When an element is broken down in alpha decay it looses two neutrons and two protons. This means hat the name of the element will change as well, moving back two places on the periodic table. Alpha decay is is not very penetrating because the He atoms capture electrons before traveling very far. However it is very damaging because the alpha particles can knock atoms off of molecules.Alpha decay is the most common in elements with an atomic number greater than 83.

- Beta negative decay follows the form:

$$ {}^{X}_{Z}A \longrightarrow {}^{0}_{-1}e^{-1} + {}^{X}_{Z+1}B $$

Beta particle ↲

The beta emission increases the atomic number by one by adding one proton. At the same time, one neutron is lost so the mass of the daughter isotope is the same as the parent isotope. Beta negative decay is more penetrating than alpha decay because the particles are smaller, but less penetrating than gamma decay. Beta electrons can penetrate through about 1 cm of flesh before they are brought to a halt because of electrostatic forces. Beta decay is most common in elements with a high neutron to proton ratio.

- Gamma decay follows the form:

$$ {}^{X}_{Z}A \rightarrow {}^{0}_{0}\gamma + {}^{X}_{Z}B $$

Gamma particle ↲

In gamma emission, neither the atomic number or the mass number is changed. A high energy gamma ray is given off when the parent isotope falls into a lower energy state. Gamma radiation is the most penetrating of all. These photons can pass through the body and cause damage by ionizing all the molecules in their way.

- Positron emission (also called Beta positive decay) follows the form:

$$ {}^{X}_{Z}A \rightarrow {}^{0}_{1}e+ \quad {}^{X}_{Z-1}B $$

Positron particle ↲

In this reaction a positron is emitted. A positron is exactly like an electron in mass and charge force except with a positive charge. It is formed when a proton breaks into a neutron with mass and no charge and this positron with no mass and the positive charge. Positron emission is most common in lighter elements with a low neutron to proton ratio.

- Electron capture follows the form:

$$ {}^{X}_{Z}A + {}^{0}_{-1}e \rightarrow {}^{X}_{Z-1}B $$

Electron particle ↲

In this reaction a nucleus captures one of its own atom's inner shell electrons which reduces the atomic number by one. This captured electron joins with a proton in the nucleus to form a neutron. Electron capture is common in larger elements with a low neutron to proton ratio.

All elements with an atomic number over 83 are considered radioactive. Radioactivity can be measured using a geiger counter, a cylinder containing a low-pressure gas and two electrodes. Radiation ionizes the atoms in the cylinder and allows current to flow between the electrodes.

All radioactive elements disintegrate according to their specific half life. The half life of a radioactive substance is the time required for half of the initial number of nuclei to disintegrate. The decay rate expresses the speed at which a substance disintegrates. The following equation represents the relationship between the number of nuclei remaining, N, the number of nuclei initially present, N_0, the rate of decay, k, and the amount of time, t.

$$ \ln \frac{N}{N_0} = -kt $$

The relationship between the half-life of a radioactive substance and k, the rate at which it decays can also be found.

$$ t_{1/2} = \frac{.693}{k} $$

By using these equations, it is possible to calculate how

much of a nuclear substance will be left after a certain time and how much of a substance originally existed. A common example is isotopic dating in which the ages of archeological artifacts are determined by measuring the activity of the isotopes.

RADIOACTIVITY: HISTORICAL FIGURES

This article will focus on the efforts of four scientists: Wilhelm Conrad Roentgen, Antoine Henri Becquerel, Marie Sklodowska Curie, and Ernest Rutherford. It emphasizes their contributions to the elucidation of radioactivity and the "key" experiments they performed pertaining to their discoveries. The biographies and photographs are adapted from The Health Physics Society Centennial Calendar by permission of the Health Physics Society.

Wilhelm Conrad Roentgen (1845-1923)

On November 8, 1895, at the University of Wurzburg, Wilhelm Roentgen's attention was drawn to a glowing fluorescent screen on a nearby table. Roentgen immediately determined that the fluorescence was caused by invisible rays originating from the partially evacuated glass Hittorf-Crookes tube he was using to study cathode rays (i.e., electrons). Surprisingly, these mysterious rays penetrated the opaque black paper wrapped around the tube. Roentgen had discovered X rays, a momentous event that instantly revolutionized the field of physics and medicine.

However, prior to his first formal correspondence to the University Physical-Medical Society, Roentgen spent two months thoroughly investigating the properties of X rays. Silvanus Thompson complained that Roentgen left "little for others to do beyond elaborating his work." For his discovery, Roentgen received the first Nobel Prize in physics in 1901. When later asked what his thoughts were at the moment of his discovery, he replied "I didn't think, I investigated. "It was the crowning achievement in a career beset by more than its share of difficulties. As a student in Holland, Roentgen was expelled from the Utrecht Technical School for a prank committed by another student. Even after receiving a doctorate, his lack of a diploma initially prevented him from obtaining a position at the University of Wurzburg. He even was accused of having stolen the discovery of X rays by those who failed to observe them.

Nevertheless, Roentgen was a brilliant experimentalist who never sought honors or financial profit for his research. He rejected a title (i.e., von Roentgen) that would have provided entry into the German nobility, and donated the money he received from the Nobel Prize to his University. Roentgen did accept the honorary degree of Doctor of Medicine offered to him by the medical faculty of his own University of Wurzburg. However, he refused to take out any patents in order that the world could freely benefit from his work. At the time of his death, Roentgen was nearly bankrupt from the inflation that followed World War I.

Antoine Henri Becquerel (1852-1908)

Henri Becquerel was born into a family of scientists. His grandfather had made important contributions in the field of electrochemistry while his father had investigated the phenomena of fluorescence and phosphorescence. Becquerel not only inherited their interest in science, he also inherited the minerals and compounds studied by his father.And so, upon learning how Wilhelm Roentgen discovered X rays from the fluorescence they produced, Becquerel had a ready source of fluorescent materials with which to pursue his own investigations of these mysterious rays.

The material Becquerel chose to work with was potassium uranyl sulfate, $K_2UO_2(SO_4)_2$, which he exposed to sunlight and placed on photographic plates wrapped in black paper. When developed, the plates revealed an image of the uranium crystals. Becquerel concluded "that the phosphorescent substance in question emits radiation which penetrates paper opaque to light." Initially he believed that the sun's energy was being absorbed by the uranium which then emitted X rays.

Further investigation, on the 26th and 27th of February, was delayed because the skies over Paris were overcast and the uranium-covered plates Becquerel intended to expose to the sun were returned to a drawer. On the first of March, he developed the photographic plates expecting only faint images to appear. To his surprise, the images were clear and strong. This meant that the uranium emitted radiation without an external source of energy such as the sun. Becquerel had discovered radioactivity, the spontaneous emission of radiation by a material.

Later, Becquerel demonstrated that the radiation emitted by uranium shared certain characteristics with X rays but, unlike X rays, could be deflected by a magnetic field and therefore must consist of charged particles. For his discovery of radioactivity, Becquerel was awarded the 1903 Nobel Prize for physics.

Pierre Curie (1859-1906)

Marie Curie (1867-1934)

By the time he met Marie Sklodowska, Pierre Curie had already established an impressive reputation. In 1880, he and his brother Jacques had discovered piezoelectricity whereby physical pressure applied to a crystal resulted in the creation of an electric potential. He also had made important investigations into the phenomenon of magnetism including the identification of a temperature, the curie point, above which a material's magnetic properties disappear. However, shortly after his marriage to Marie in 1895, Pierre subjugated his research to her interests.

Together, they began investigating the phenomenon of radioactivity recently discovered in uranium ore. Although the phenomenon was discovered by Henri Becquerel, the term radioactivity was coined by Marie. After chemical extraction of uranium from the ore, Marie noted the residual material to be more "active" than the pure uranium.

She concluded that the ore contained, in addition to uranium, new elements that were also radioactive. This led to their discoveries of the elements of polonium and radium, but it took four more years of processing tons of ore under oppressive conditions to isolate enough of each element to determine its chemical properties. For their work on radioactivity, the Curies were awarded the 1903 Nobel Prize in physics. Tragically, Pierre was killed three years later in an accident while crossing a street in a rainstorm. Pierre's teaching position at the Sorbonne was given to Marie.

Never before had a woman taught there in its 650 year history! Her first lecture began with the very sentence her husband had used to finish his last. In his honour, the 1910 Radiology Congress chose the curie as the basic unit of radioactivity: the quantity of radon in equilibrium with one gram of radium (current definition: 1 Ci = 3.7x1010 dps).

A year later, Marie was awarded the Nobel Prize in chemistry for her discoveries of radium and polonium, thus becoming the first person to receive two Nobel Prizes. For the remainder of her life she tirelessly investigated and promoted the use if radium as a treatment for cancer. Marie Curie died July 4, 1934, overtaken by pernicious anemia no doubt caused by years of overwork and radiation exposure.

Ernest Rutherford (1871-1937)

Ernest Rutherford is considered the father of nuclear physics. Indeed, it could be said that Rutherford invented the very language to describe the theoretical concepts of the atom and the phenomenon of radioactivity. Particles named and characterized by him include the alpha particle, beta particle and proton.

Even the neutron, discovered by James Chadwick, owes its name to Rutherford. The exponential equation used to calculate the decay of radioactive substances was first employed for that purpose by Rutherford and he was the first to elucidate the related concepts of the half-life and decay constant. With Frederick Soddy at McGill University, Rutherford showed that elements such as uranium and thorium became different elements (i.e., transmuted) through the process of radioactive decay.

At the time, such an incredible idea was not to be mentioned in polite company: it belonged to the realm of alchemy, not science. For this work, Rutherford won the 1908 Nobel Prize in chemistry. In 1909, now at the University of Manchester, Rutherford was bombarding a thin gold foil with alpha particles when he noticed that although almost all of them went through the gold, one in eight thousand would "bounce" (i.e., scatter) back. The amazed Rutherford commented that it was "as if you fired a 15-inch naval shell at a piece of tissue paper and the shell came right back and hit you."

From this simple observation, Rutherford concluded that the atom's mass must be concentrated in a small positively-charged nucleus while the electrons inhabit the farthest reaches of the atom. Although this planetary model of the atom has been greatly refined over the years, it remains as valid today as when it was originally formulated by Rutherford.

In 1919, Rutherford returned to Cambridge to become director of the Cavendish laboratory where he had previously done his graduate work under J.J. Thomson. It was here that he made his final major achievement, the artificial alteration of nuclear and atomic structure. By bombarding nitrogen with alpha particles, Rutherford demonstrated the production of a different element, oxygen.

"Playing with marbles" is what he called; the newspapers reported that Rutherford had "split the atom." After his death in 1937, Rutherford's remains were buried in Westminster Abbey near those of Sir Isaac Newton.

Chapter 2

Nuclear Structure

Understanding the structure of the atomic nucleus is one of the central problems in nuclear physics..

THE LIQUID DROP MODEL

This is one of the first models of nuclear structure, proposed by Carl Friedrich von Weizsäcker in 1935. It describes the nucleus as a classical fluid made up of neutrons and protons, with an internal repulsive electric force proportional to the number of protons.

The quantum mechanical nature of these particles appears via the Pauli exclusion principle, which states that no two nucleons of the same kind can be at the same state. Thus the fluid is actually what is known as a fermi fluid. This simple model reproduces the main features of the binding energy of nuclei.

THE SHELL MODEL

Systematic measurements of the binding energy of atomic nuclei show systematic deviations with respect to those estimated from the liquid drop model. In particular, some nuclei having special values for their number of protons and/or neutrons are much more bound than predicted by the liquid drop model. These nuclei are called simply/doubly magic.

This observation led scientists to assume the existence of a shell structure of nucleons (protons and neutrons) within the nucleus, like that of electrons within atoms. Indeed nucleons are quantum objects. Strictly speaking, one should not speak of energies of individual nucleons, because they are all correlated with each other. To be able to speak of a shell structure, one envisions first an average nucleus, within which nucleons propagate individually. Owing to their quantum character, they can have then only *discrete* values of energy levels.

These levels are by no means uniformly distributed : some intervals of energy are crowded, but they are separated by almost empty gaps. A shell is such a set of levels separated from the other ones by a wide empty gap.

The determination of the energy levels is done via quantum mechanics, more precisely by diagonalization of the single-nucleon Hamiltonian. Each level may be occupied by a nucleon, or empty. Some levels accommodate several different quantum states with the same energy : they are said to be *degenerate*. This occurs in particular if the average nucleus has some symmetry.

The concept of shells allows one to understand why some nuclei are more bound than others. This is because two nucleons of the same kind cannot be in the same state (Pauli exclusion principle). So the lowest-energy state of the nucleus is one where nucleons fill all energy levels from the bottom up to some level. A nucleus with full shells is exceptionally stable, as will be explained.

As with electrons in the periodic table, protons in the outermost shell are relatively loosely bound to the nucleus if there are only few protons in that shell, because they are farthest from the centre of the nucleus.

Therefore nuclei which have a full outer proton shell will be more tightly bound and have a higher binding energy than other nuclei with a similar total number of protons. All this is also true for neutrons. The binding is said to be "magically" large.

Furthermore, the energy needed to excite the nucleus (i.e. moving a nucleon to a higher, previously unoccupied level) is exceptionally high in such nuclei. Whenever this unoccupied level is the next after a full shell, the only way to excite the nucleus is to raise one nucleon *across the gap*, thus spending a large amount of energy. Otherwise, if the highest occupied energy level lies in a partly filled shell, much less energy is required to raise a nucleon to a higher state in the same shell.

Basic Hypotheses

The expression "shell model" is ambiguous in that it refers to two different eras in the state of the art. It was previously used to describe the existence of nucleon shells in the nucleus according to an approach closer to what is now called mean field theory. Nowadays, it refers to a set of techniques which help solving some variants of the nuclear *n*-body problem. We shall introduce these here.

Several basic hypotheses are made in order to give a precise conceptual framework to the shell model :

- The atomic nucleus is a quantum *n*-body system.
- The nucleus is not a relativistic object. The equation of motion giving the system wavefunction (which contains the whole of the information about it in quantum mechanics), is the Schrödinger equation (which is non-relativistic).
- The nucleons interact only via a 2-body interaction. This limitation is in effect a practical consequence of the Pauli exclusion principle : the mean free path of a nucleon being large with respect to the nucleus size, the probability that three nucleons interact simultaneously is considered as small enough to be negligible.
- Nucleons are considered to be pointlike, without any structure, in this model, for the sake of simplicity.

Brief Description of the Formalism

The general process used in the shell model calculations is the following. First a Hamiltonian for the nucleus is defined. Only 1- and 2-body terms are taken into account in this definition. The interaction is an effective theory : it contains free parameters which have to be fitted with experimental data.

The next step consists in defining a basis of single-particle states, i.e. a set of wavefunctions describing all possible nucleon states. Most of the time, this basis is obtained via a

Hartree-Fock computation. With this set of 1-particle states, Slater determinants are built, that is, wavefunctions for Z proton variables or N neutron variables, which are antisymmetrized products of single-particle wavefunctions (antisymmetrized meaning that under exchange of variables for any pair of nucleons, the wavefunction only changes sign).

In principle, the number of quantum states available for a single nucleon at a finite energy is finite, say n. The number of nucleons in the nucleus must be smaller than the number of available states, otherwise the nucleus cannot hold all of its nucleons.

There are thus several ways to choose Z (or N) states among the n possible. In combinatorial mathematics, the number of choices of Z objects among n is the binomial coefficient C_n^Z. If n is much larger than Z (or N), this increases roughly like n^Z. Practically, this number becomes so large that every computation is impossible for $A=N+Z$ larger than 8.

To obviate this difficulty, the space of possible single-particle states is divided into a core and a valence shell, by analogy with chemistry. The core is a set of single-particles which are assumed to be inactive, in the sense that they are the well bound lowest-energy states, and that there is no need to reexamine their situation.

They do not appear in the Slater determinants, contrary to the states in the valence space, which is the space of all single-particle states *not in the core*, but possibly to be considered in the choice of the build of the (Z-) N-body wavefunction. The set of all possible Slater determinants in the valence space defines a basis for (Z-) N-body states.

The last step consists in computing the matrix of the Hamiltonian within this basis, and to diagonalize it. In spite of the reduction of the dimension of the basis owing to the fixation of the core, the matrices to be diagonalized reach easily dimensions of the order of 10^9, and demand specific diagonalization techniques.

The shell model calculations give in general an excellent fit with experimental data. They depend however strongly on two main factors :

- The way to divide the single-particle space into core and valence.
- The effective nucleon-nucleon interaction.

The Independent-particle Model

The interaction between nucleons, which is a consequence of strong interactions, and binds the nucleons within the nucleus exhibits the peculiar behaviour to have a finite range : it vanishes when the distance among two nucleons becomes too large ; it is attractive at medium range, and repulsive at very small range.

This last property correlates with the Pauli exclusion principle according to which two fermions (nucleons are fermions) cannot be in the same quantum state. This results in a very large mean free path of a nucleon within the nucleus. This result, confirmed by particle scattering experiments, led to the elaboration of the independent-particle model.

The main idea of this approach is that all looks as if a nucleon moves inside a certain potential well (which keeps it bound to the nucleus) independently from the other nucleons. In theory, this amounts to replacing a *N*-body problem (*N* particles interacting) by *N* single-body problems. This essential simplification of the problem is the cornerstone of mean field theories. These are also widely used in atomic physics, where electrons move in a mean field due to the central nucleus and the electron cloud itself.

Although this hypothesis looks grossly simplifying, it led to big success, and mean field theories are now a basic part of atomic nucleus theory. One should also notice that they are modular enough (in terms of programming theory), in that it is quite easy to introduce certain effects such as nucleon pairing, or collective motions of the nucleon like rotation, or vibration, adding "by hand" the corresponding energy terms in the formalism.

Nuclear Potential and Effective Interaction

A large part of the practical difficulties met in mean field theories is the definition (or calculation) of the potential of the mean field itself. One can very roughly distinguish between two approaches :

- The phenomenological approach is a parameterization of the nuclear potential by an appropriate mathematical function. Historically, this procedure was applied with the greatest success by Sven Gösta Nilsson, who used as a potential a (deformed) harmonic oscillator potential. The most recent parameterizations are based on more realistic functions, which account more accurately for scattering experiments.
- The self-consistent or Hartree-Fock approach aims to deduce mathematically the nuclear potential from the nucleon-nucleon interaction. This technique implies a resolution of the Schrödinger equation in an iterative fashion, since the potential depends there upon the wavefunctions to be determined. The latter are written as Slater determinants.

In the case of the Hartree-Fock approaches, the trouble is not to find the mathematical function which describes best the nuclear potential, but that which describes best the nucleon-nucleon interaction. Indeed, in contrast with atomic physics where the interaction is known (it is the Coulomb interaction), the nucleon-nucleon interaction within the nucleus is not known analytically.

There are two main reasons for this fact. First, the strong interaction acts essentially among the quarks forming the nucleons. The nucleon-nucleon interaction in vacuum is a mere *consequence* of the quark-quark interaction.

While the latter is well understood in the framework of the standard model at high energies, it is much more complicated in low energies due to Colour confinement and asymptotic freedom. Thus there is no fundamental theory allowing one to deduce the nucleon-nucleon interaction from the quark-quark interaction.

Further, even if this problem were solved, there would remain a large difference between the ideal (and conceptually simpler) case of two nucleons interacting in vacuo, and that of these nucleons interacting in the nuclear matter. To go further, it was necessary to invent the concept of effective interaction. The latter is basically a mathematical function with several arbitrary parameters, which are adjusted to agree with experimental data.

The Self-consistent Approaches of the Hartree-Fock Type

In the Hartree-Fock approach of the *n*-body problem, the starting point is a Hamiltonian containing *n* kinetic energy terms, and potential terms. One of the mean field theory hypotheses is that only the 2-body interaction is to be taken into account. The potential term of the Hamiltonian represents all possible 2-body interactions in the set of *n* fermions. It is the first hypothesis.

The second step consists in assuming that the wavefunction of the system can be written as a Slater determinant. This statement is the mathematical translation of the independent-particle model. This is the second hypothesis.

There remains now to determine the components of this Slater determinant, that is, the individual wavefunctions of the nucleons. To this end, it is assumed that the total wavefunction (the Slater determinant) is such that the energy is minimum. This is the third hypothesis.

Technically, it means that one must compute the mean value of the (known) 2-body Hamiltonian on the (unknown) Slater determinant, and impose that its mathematical variation vanishes. This leads to a set of equations where the unknowns are the individual wavefunctions : the Hartree-Fock equations. Solving these equations gives the wavefunctions and individual energy levels of nucleons, and so the total energy of the nucleus and its wavefunction.

This short account of the Hartree-Fock method explains why it is called also the variational approach. At the beginning of the calculation, the total energy is a "function of the individual wavefunctions" (a so-called functional), and everything is then made in order to optimize the choice of these wavefunctions so that the functional has a minimum –

hopefully absolute, and not only local. To be more precise, there should be mentioned that the energy is a functional of the density, defined as the sum of the individual squared wavefunctions. Let us note also that the Hartree-Fock method is also used in atomic physics and condensed matter physics as Density Functional Theory, DFT.

The process of solving the Hartree-Fock equations can only be iterative, since these are in fact a Schrödinger equation in which the potential depends on the density, that is, precisely on the wavefunctions to be determined. Practically, the algorithm is started with a set of individual grossly reasonable wavefunctions (in general the eigenfunctions of a harmonic oscillator).

These allow to compute the density, and therefrom the Hartree-Fock potential. Once this done, the Schrödinger equation is solved anew, and so on. The calculation stops – convergence is reached – when the difference among wavefunctions, or energy levels, for two successive iterations is less than a fixed value.

Then the mean field potential is completely determined, and the Hartree-Fock equations become standard Schrödinger equations. The corresponding Hamiltonian is then called the Hartree-Fock Hamiltonian.

The Relativistic Mean Field Approaches

Born first in the 1970's with the works of D. Walecka on quantum hadrodynamics, the relativistic models of the nucleus were sharpened up towards the end of the 1980's by P. Ring and coworkers. The starting point of these approaches is the relativistic quantum field theory. In this context, the nucleon interactions occur via the exchange of virtual particles called mesons.

The idea is, in a first step, to build a lagrangian containing these interaction terms. Second, by an application of the least action principle, one gets a set of equations of motion. The real particles (here the nucleons) obey the Dirac equation, whilst the virtual ones (here the mesons) obey the Klein-Gordon equations.

In view of the non-perturbative nature of strong interaction, and also of the fact that it is relatively badly known, the use of such an approach in the case of atomic nuclei requires drastic approximations.

The main simplification consists in replacing in the equations all field terms (which are operators in the mathematical sense) by their mean value (which are functions). In this way, one gets a system of coupled integro-differential equations, which can be solved numerically, if not analytically.

Spontaneous Breaking of Symmetry in Nuclear Physics

One of the focal points of all physics is symmetry. The nucleon-nucleon interaction and all effective interactions used in practice have certain symmetries. They are invariant by translation (changing the frame of reference so that directions are not altered), by rotation (turning the frame of reference around some axis), or parity (changing the sense of axes) in the sense that the interaction does not change under any of these operations.

Nevertheless, in the Hartree-Fock approach, solutions which are not invariant under such a symmetry can appear. One speaks then of spontaneous symmetry breaking.

Qualitatively, these spontaneous symmetry breakings can be explained in the following way : in the mean field theory, the nucleus is described as a set of independent particles. Most additional correlations among nucleons which do not enter the mean field are neglected.

They can appear however by a breaking of the symmetry of the mean field Hamiltonian, which is only approximate. If the density used to start the iterations of the Hartree-Fock process breaks certain symmetries, the final Hartree-Fock Hamiltonian may break these symmetries, if it is advantageous to keep these broken from the point of view of the total energy.

It may also converge towards a symmetric solution. In any case, if the final solution breaks the symmetry, for example, the rotational symmetry, so that the nucleus appears not to be spherical, but elliptic, all configurations deduced from this deformed nucleus by a rotation are just as good solutions for the Hartree-Fock problem. The ground state of the nucleus is then *degenerate*.

A similar phenomenon happens with the nuclear pairing, which violates the conservation of the number of baryons.

Nuclear Pairing Phenomenon

Historically, the observation that the nuclei with an even number of nucleons are systematically more bound than those with an odd one led to propose the nuclear pairing hypothesis. The very simple idea is that each nucleon binds with another one to form a pair.

When the nucleus has an even number of nucleons, each one of them finds a partner. To excite such a system, one must at least use such an energy as to break a pair. Conversely, in the case of odd number of nucleons, there exists a bachelor nucleon, which needs less energy to be excited.

This phenomenon is closely analogous to that of superconductivity in solid state physics (at least the low-temperature superconductivity). The first theoretical description of nuclear pairing was proposed at the end of the 1950's by Aage Bohr and Ben Mottelson. It was close to the BCS theory of Bardeen, Cooper and Schrieffer, which accounts for metal superconductivity. Theoretically, the pairing phenomenon as described by the BCS theory combines with the mean field theory : nucleons are both subject to the mean field potential and to the pairing interaction, but these are independent.

It is tempting to interpret the pairing interaction as a residual interaction. To build the mean field interaction, only some terms of the nucleon-nucleon interaction are taken into account. Everything else is qualified as *residual interaction*. The value of the mean field theory rests on the fact that the *residual interaction* is numerically much less than what is taken into account for the mean field.

There should be however a link between them, since they proceed from the same nucleon-nucleon interaction. This is not taken into account in the BCS theory. To circumvent this point, the Hartree-Fock-Bogolyubov (HFB) approach has been developed, to include into a unified formalism the mean field, the pairing and their mutual links.

Let us note finally that one big difference between superconductivity and nuclear pairing resides in the number of particles. In a metal, the number of free electrons is very large, compared to the number of nucleons in a nucleus. The BCS (and HFB) approach describes the system wave function as a superposition of components with different numbers of particles.

In the case of a metal, this violation of the conservation of the number of particles is of no consequence, in view of the huge statistics.

But in nuclear physics, this leads to a real problem. Specific techniques for restoring the number of particles have been developed, in the framework of the restoration of broken symmetries.

Chapter 3

Atomic Nucleus

The nucleus of an atom is the very dense region, consisting of nucleons (protons and neutrons), at the centre of an atom. Although the size of the nucleus varies considerably according to the mass of the atom, the size of the entire atom is comparatively constant.

Almost all of the mass in an atom is made up from the protons and neutrons in the nucleus with a very small contribution from the orbiting electrons. The diameter of the nucleus is in the range of 1.6 fm (1.6×10^{-15} m) (for a proton in light hydrogen) to about 15 fm (for the heaviest atoms, such as uranium). These dimensions are much smaller than the size of the atom itself by a factor of about 23,000 (uranium) to about 145,000 (hydrogen). The branch of physics concerned with studying and understanding the atomic nucleus, including its composition and the forces which bind it together, is called nuclear physics.

ETYMOLOGY

The term nucleus is from Latin *nucleus* ("'kernel'"), derived from *nux* ("'nut'"). In 1844, Michael Faraday used the term to refer to the "central point of an atom". The modern atomic meaning was proposed by Ernest Rutherford in 1912. The adoption of the term "nucleus" to atomic theory, however, was not immediate.

In 1916, for example, Gilbert N. Lewis stated, in his famous article *The Atom and the Molecule,* that "the atom is composed of the *kernel* and an outer atom or *shell*".

A figurative depiction of the helium-4 atom. In the

nucleus, the two protons are shown in red and neutrons blue. This depiction shows the particles as separate, whereas in an actual helium atom, the protons are superimposed in space and most likely found at the very centre of the nucleus, and the same is true of the two neutrons.

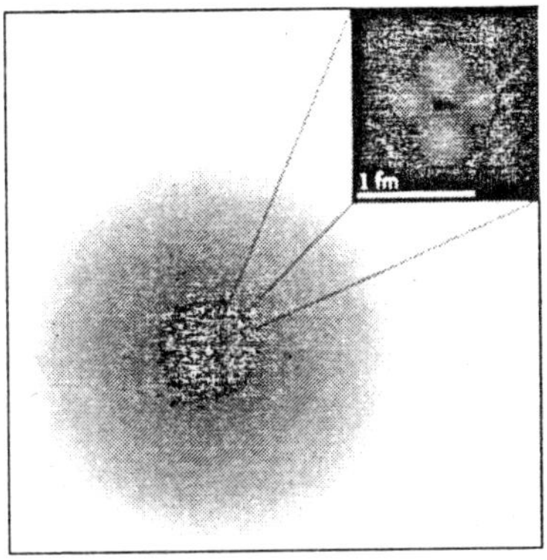

Fig. Helium-4 Atom.

Thus all four particles are most likely found in exactly the same space. Classical images of separate particles thus fail to model known charge distributions in very small nuclei

Nuclear Makeup

The nucleus of an atom consists of protons and neutrons (two types of baryons) bound by the nuclear force (also known as the residual strong force). These baryons are further composed of subatomic fundamental particles known as quarks bound by the strong interaction. Which chemical element an atom represents is determined by the number of protons in the nucleus. Each proton carries a single positive charge, and the total electrical charge of the nucleus is spread fairly uniformly throughout its body, with the exception of hydrogen and helium, where the charge is concentrated most highly at the central point, as would be expected for fermions in 1s states without orbital angular momentum.

As each proton carries a unit of charge, the charge distribution is indicative of the proton distribution, and the neutron distribution probably is similar.

PROTONS AND NEUTRONS

Protons and neutrons are fermions, with different values of the isospin quantum number, so two protons and two neutrons can share the same space wave function. They sometimes are viewed as two different states of the same particle, the *nucleon*.

In the rare case of a hypernucleus, a third baryon called a hyperon, with a different value of the strangeness quantum number can also share the wave function. However, the latter type of nuclei are extremely unstable and are not found on Earth except in high energy physics experiments.

The neutron has a positively charged core of radius ≈ 0.3 fm surrounded by a compensating negative charge of radius between 0.3 fm and 2 fm. The proton has an approximately exponentially decaying charge distribution with a mean square radius of about 0.8 fm.

FORCES

Nuclei are bound together by the residual strong force. The residual strong force is minor residuum of the strong force which binds quarks together to form protons and neutrons. This force is much weaker *between* neutrons and protons because it is mostly neutalized within them, in the same way that electromagnetic forces between neutral atoms (van der Waals forces) are much weaker than the electromagnetic forces that hold the atoms internally together.

The strong force is so named because it is significantly larger in magnitude than the other fundamental forces (electroweak, electromagnetic and gravitational). The strong force is highly attractive at very small distances, and this overwhelms the repulsion between protons due to the electromagnetic force, thus allowing nuclei to exist. However, because the residual strong force has a limited range because it decays quickly with distance, only nuclei smaller than a certain size can be completely stable.

The largest known complete stable nucleus is lead-208 which contains 208 neutrons and protons. Nuclei larger than this maximal size of 208 particles generally become increasingly short-lived, as the number of neutrons and protons which compose them increases beyond this number.

The residual strong force usually acts over a very short range (a few fermis, roughly one or two nucleon diameters) and causes an attraction between nucleons (protons and neutrons). However there are also halo nuclei such as lithium-11 or boron-14, in which dineutrons, or other collections of nucleons, orbit at distances of about ten fermis (similar to the size of lead-208). Such nuclei are always short-lived; for example, lithium-11 has a half-life of less than 8.6 milliseconds.

NUCLEAR SIZE

The nuclear radius (R) is considered to be one of the basic things that any model must explain. It is roughly proportional to the cube root of the mass number (A) of the nucleus, and particularly in nuclei containing many nucleons, as they arrange in more spherical configurations:

The nucleus has approximately a constant density and therefore the nuclear radius R can be approximated by the following formula,

$$R = r_0 A^{1/3}$$

where, A = Atomic mass number (the number of protons, Z, plus the number of neutrons, N) and $r_0 = 1.25$ fm $= 1.25 \times 10^{-15}$ m. In this equation, the constant r_0 varies by 0.2 fm, depending on the nucleus in question, but this is less than 20 per cent change from a constant.

In other words, packing protons and neutrons in the nucleus gives *approximately* the same total size result as packing hard spheres of a constant size (like marbles) into a tight bag.

LIQUID DROP MODELS

Early models of the nucleus viewed the nucleus as a rotating liquid drop. In this model, the trade-off of long-range electromagnetic forces and relatively short-range nuclear forces, together cause behaviour which resembled surface tension forces in liquid drops of different sizes. This formula is successful at explaining many important phenomena of nuclei, such as their changing amounts of binding energy as their size and composition changes, but it does not explain the special stability which occurs when nuclei have special "magic numbers" of protons or neutrons.

SHELL MODELS AND OTHER QUANTUM MODELS

A number of models for the nucleus have also been proposed in which nucleons occupy orbitals, much like the atomic orbitals in atomic physics theory.

These wave models imagine nucleons to be either sizeless point particles in potential wells, or else probability waves as in the "optical model", frictionlessly orbiting at high speed in potential wells.

In these models, the nucleons occupy orbitals in pairs, due to being fermions, but the exact nature and capacity of nuclear shells differs somewhat from those of elections in atomic orbitals, primarily because the potential well in which the nucleons move (especially in larger nuclei) is quite different from the central electromagnetic potential well which binds electrons in atoms.

Nevertheless, the resemblance to atomic orbital models may be seen in a small atomic nucleus like that of helium-4, in which the two protons and two neutrons separately occupy 1s orbitals analogous to the 1s orbitals for the two electrons in the helium atom, and achieve unusual stability for the same reason. This stability also underlies the fact that nuclei with 5 nucleons are all extremely unstable and short-lived.

For larger nuclei, the shells occupied by nucleons begin to differ significantly from electron shells, but nevertheless, present nuclear theory does predict the magic numbers of filled nuclear shells for both protons and neutrons. The closure of the stable shells predicts unusually stable configurations, analogous to the noble group of nearly-inert gases in chemistry.

An example is the stability of the closed shell of 50 protons, which allows tin to have 10 stable isotopes, more than any other element. Similarly, the distance from shell-closure explains the unusual instability of isotopes which have far from stable numbers of these particles, such as the radioactive elements 43 (technetium) and 61 (promethium), each of which is preceded and followed by 17 or more stable elements.

NUCLEAR STRUCTURE

As with the case of superfluid liquid helium, atomic nuclei are an example of a state in which both :

- "Ordinary" particle physical rules for volume
- Non-intuitive quantum mechanical rules for a wave-like nature apply.

In superfluid helium, the helium atoms have volume, and essentially "touch" each other, yet at the same time exhibit strange bulk properties, consistent with a Bose-Einstein condensation. The latter reveals that they also have a wave-like nature and do not exhibit standard fluid properties, such as friction. For nuclei made of hadrons which are fermions, the same type of condensation does not occur, yet nevertheless, many nuclear properties can only be explained similarly by a combination of properties of particles with volume, in addition to the frictionless motion characteristic of the wave-like behaviour of objects trapped in Schroedinger quantum orbitals.

ATOMIC STRUCTURE

DISCOVERY OF ,α β & γ PARTICLES

The discovery of x-rays by William Conrad Roentgen in November of 1895 excited the imagination of a generation of scientists who rushed to study this phenomenon. Within a few months, Henri Becquerel found that both uranium metal and salts of this element gave off a different form of radiation, which could also pass through solids. By 1898, Marie Curie found that compounds of thorium were also "radioactive." After pain-staking effort she eventually isolated two more radioactive elements – polonium and radiumfrom ores that contained uranium.

In 1899 Ernest Rutherford found that there were at least two different forms of radioactivity when he studied the absorption of radioactivity by thin sheets of metal foil. One, which he called alpha (α) particles, were absorbed by metal foil that was a few hundredths of a centimeter thick. The other, beta (β) particles, could pass through 100 times as much metal foil before they became absorbed. Shortly thereafter, a third form of radiation, gamma (γ) rays, was discovered that could penetrate as much as several centimeters of lead.

The results of early experiments on these three forms of radiation are shown in the figure below. The direction in which -particles were deflected by an electric field suggested that they were positively charged. The magnitude of this deflection suggested that they had the same charge-to-mass ratio as an He^{2+} ion.

To test the equivalence between -particles and He^{2+} ions, Rutherford built an apparatus that allowed -particles to pass through a very thin glass wall into an evacuated flask that contained a pair of metal electrodes. After a few days, he connected these electrodes to a battery and noted that the gas in the flask did indeed give off the characteristic emission spectrum of helium.

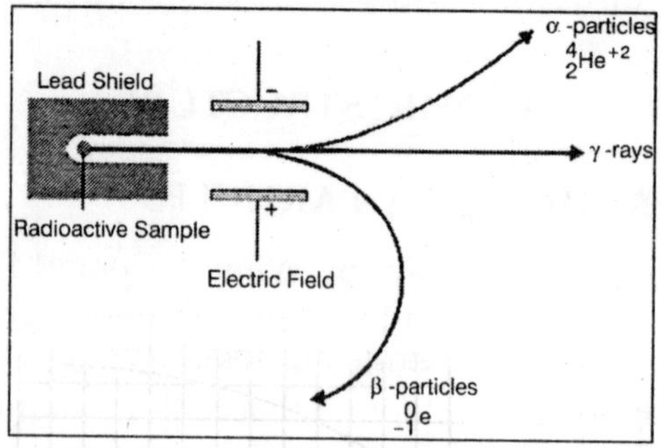

Fig. The Effect of an Electric Field on α-, β, and γ-Radiation.

Experiments with electric and magnetic fields demonstrated that β particles were negatively charged. Furthermore, they had the same charge-to-mass ratio as an electron. To date, no detectable difference has been found between β-particles and electrons. The only reason to retain the name " β-particle" is to emphasize the fact that these particles are ejected from the nucleus of an atom when it undergoes radioactive decay. The fact that -rays are not deflected by either electric or magnetic fields suggests that these rays don't carry an electric charge. Since they travel at the speed of light, they are classified as a form of electromagnetic radiation that carries even more energy than x-rays.

Properties of Radioactive Particles							
Name	*Symbols*	*Identity*	*Relative Charge*	*Relative Mass*	*Penetrating Power*	*Interaction with Charged Plates*	*Hazards*
Alpha	α, $^{4}_{2}He$	Helium nucleus	2+	4	low - stopped by a sheet of paper	attracted to negative plate, deflected by positive plate	harmful if ingested
Beta	β, $^{0}_{-1}e$	Electron	1-	$\frac{1}{2000}$	moderate – pass through paper and ½ mm aluminium, stopped by ½ mm lead	attracted to positive plate, deflected by negative plate	skin burns, harmful if ingested particularly iodine-131 in thyroid & strontium-90 in bones
Gamma	γ	Electromagnetic radiation	0	0	high - only stopped by several centimetres of lead or many centimetres of concrete	unaffected	most dangerous as these are the most penetrating, as a consequence, gamma rays can be used to sterilize materials & destroy bacteria in food

GROWTH AND DECAY CURVES

At the turn of the century, when radioactivity was discovered, atoms were assumed to be indestructible. Ernest Rutherford and Frederick Soddy, however, found that radioactive substances became less active with time, as shown in the figure below. More importantly, they noticed that radioactivity was always accompanied by the formation of atoms of a different element. By 1903, they concluded that radioactivity was accompanied by a change in the structure of the atom. They therefore assumed that radiation was emitted when an element decayed into a different kind of atom.

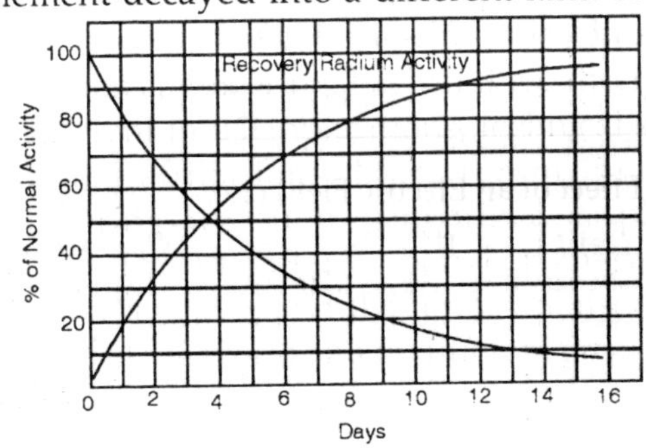

Fig. Growth and Decay Curves Reported by Rutherford and Soddy for "Uranium X" Produced when Uranium Undergoes Radioactive Decay

By 1910, 40 radioactive elements had been isolated that were associated with the process by which uranium metal decayed to lead. This created a problem, however, because there was space for only 11 elements between lead and uranium. In 1913, Kasimir Fajans and Fredick Soddy proposed an explanation for these results based on the following rules.

- α -particles are emitted when an element is formed that belongs two spaces to the left in the periodic table. Uranium (Z = 92), for example, emits an α -particle when it decays to form thorium (Z = 90).
- β -particles are emitted when an element is formed that belongs one space to the right in the periodic table. Actinium (Z = 89) emits a β -particle when it decays to form thorium (Z = 90).
- Radioactive elements that fall in the same place in the periodic table are different forms of the same element. The radioactive thorium produced by the α-particle decay of uranium is a different form of the element than the radioactive thorium obtained by the β -particle decay of actinium.

Soddy proposed the name isotope to describe different radioactive atoms that occupy the same position in the periodic table. J. J. Thomson and Francis Aston then used a mass spectrometer to show that isotopes are atoms of the same element that have different atomic masses.

THE STRUCTURE OF THE ATOM

The discovery of the electron in 1897 by J. J. Thomson suggested that there was an internal structure to the "indivisible" building blocks of matter known as atoms. This raised an obvious question: How many electrons does an atom contain? By studying the scattering of light, x-rays, and -particles, Thomson concluded that the number of electrons in an atom was between 0.2 and 2 times the weight of the atom.

In 1911, Rutherford concluded that the scattering of -particles by extremely thin pieces of metal foil could be explained by assuming that all of the positive charge and most of the mass of the atom were concentrated in an infinitesimally small fraction of the total volume of the atom, for which he proposed the name nucleus. Rutherford's data also suggested that the nucleus of a gold atom carries a positive charge that is about 80 times the charge on an electron.

The discovery of the neutron in 1932 explained the discrepancy between the charge on the nucleus and the mass of an atom. A neutral gold atom that has a mass of 197 amu consists of a nucleus that contains 79 protons and 118 neutrons surrounded by 79 electrons. By convention, this information is specified by the following symbol, which describes the only naturally occurring isotope of gold.

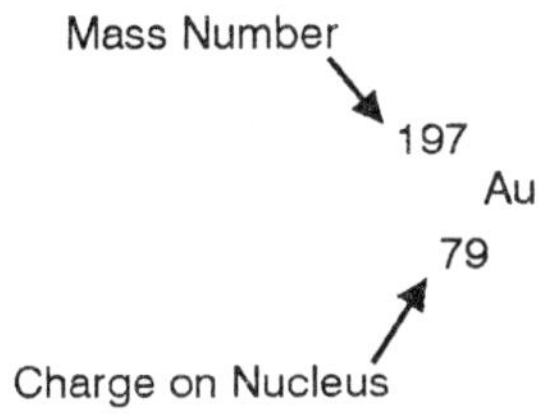

This convention can also be applied to subatomic particles. The only difference is the use of lowercase letters to identify the particle. Because anyone with access to a periodic table can find the atomic number of an element, a shorthand notation is often used that reports only the mass number of the atom and the symbol of the element. The shorthand notation for the naturally occurring isotope of gold is ^{197}Au.

RADIOACTIVE DECAY

α β and Radiation

Early studies of radioactivity indicated that three different kinds of radiation were emitted, symbolized by the first three letters of the Greek alphabet, β and. With time, it became apparent that this classification scheme was much too simple. The emission of a negatively charged β particle, for example, is only one example of a family of radioactive transformations known as β -decay. A fourth category, known as spontaneous fission, also had to be added to describe the process by which certain radioactive nuclides decompose into fragments of different weight.

ALPHA DECAY

Alpha decay is usually restricted to the heavier elements in the periodic table. (Only a handful of nuclides with atomic numbers less than 83 emit an -particle.) The product of -decay is easy to predict if we assume that both mass and charge are conserved in nuclear reactions. Alpha decay of the ^{238}U "parent" nuclide, for example, produces ^{234}Th as the "daughter" nuclide.

$$^{238}_{92}U \rightarrow ^{234}_{90}Th + ^{4}_{2}He \quad (\alpha\text{-decay})$$

The sum of the mass numbers of the products (234 + 4) is equal to the mass number of the parent nuclide (238), and the sum of the charges on the products (90 + 2) is equal to the charge on the parent nuclide.

BETA DECAY

There are three different modes of beta decay:

- electron (b^-) emission
- electron capture
- positron (b^+) emission

Electron (β^-) emission is literally the process in which an electron is ejected or emitted from the nucleus. When this happens, the charge on the nucleus increases by one. Electron (β^-) emitters are found throughout the periodic table, from the lightest elements (^3H) to the heaviest (^{255}Es).

The product of β^--emission can be predicted by assuming that both mass number and charge are conserved in nuclear reactions. If ^{40}K is a β^--emitter, for example, the product of this reaction must be ^{40}Ca.

$$^{40}_{19}K \rightarrow ^{40}_{20}Ca + ^{0}_{-1}e \quad \left[electron(\beta^-)emission\right]$$

Once again the sum of the mass numbers of the products is equal to the mass number of the parent nuclide and the sum of the charge on the products is equal to the charge on the parent nuclide. Nuclei can also decay by capturing one of the electrons that surround the nucleus. Electron capture leads to a decrease of one in the charge on the nucleus.

The energy given off in this reaction is carried by an x-ray photon, which is represented by the symbol hv, where h is Planck's constant and v is the frequency of the x-ray. The product of this reaction can be predicted, once again, by assuming that mass and charge are conserved.

$$ {}^{40}_{19}K + {}^{0}_{20}e \rightarrow {}^{40}_{-1}Ar + h\nu \quad (electron\ capture) $$

The electron captured by the nucleus in this reaction is usually a 1s electron because electrons in this orbital are the closest to the nucleus. A third form of beta decay is called positron (β^+) emission. The positron is the antimatter equivalent of an electron. It has the same mass as an electron, but the opposite charge. Positron (β^+) decay produces a daughter nuclide with one less positive charge on the nucleus than the parent.

$$ {}^{40}_{19}K \rightarrow {}^{40}_{18}Ar + {}^{0}_{+1}e \quad [positron(\beta^+)emission] $$

Positrons have a very short life-time. They rapidly lose their kinetic energy as they pass through matter. As soon as they come to rest, they combine with an electron to form two γ-ray photons in a matter-antimatter annihilation reaction.

$$ {}^{0}_{+1}e + {}^{0}_{-1}e \rightarrow 2\gamma $$

Thus, although it is theoretically possible to observe a fourth mode of beta decay corresponding to the capture of a positron, this reaction does not occur in nature. Note that in all three forms of β -decay for the ${}^{40}K$ nuclide the mass number of the parent and daughter nuclides are the same for electron emission, electron capture, and position emission. All three forms of β -decay therefore interconvert isobars.

GAMMA EMISSION

The daughter nuclides produced by α-decay or β-decay are often obtained in an excited state. The excess energy associated with this excited state is released when the nucleus emits a photon in the γ-ray portion of the electromagnetic spectrum.

Most of the time, the γ-ray is emitted within 10^{-12} seconds after the α-particle or β-particle. In some cases, gamma decay is delayed, and a short-lived, or metastable, nuclide is formed, which is identified by a small letter m written after the mass number. ^{60m}Co, for example, is produced by the electron emission of ^{60}Fe.

$$^{60}_{26}Fe \rightarrow \,^{60m}_{27}Co + \,^{0}_{-1}e$$

The metastable ^{60m}Co nuclide has a half-life of 10.5 minutes. Since electromagnetic radiation car ries neither charge nor mass, the product of γ-ray emission by ^{60m}Co is ^{60}Co.

$$^{60m}_{27}Co \rightarrow \,^{60}_{27}Co + \gamma \quad (\gamma\text{-ray emission})$$

SPONTANEOUS FISSION

Nuclides with atomic numbers of 90 or more undergo a form of radioactive decay known as spontaneous fission in which the parent nucleus splits into a pair of smaller nuclei. The reaction is usually accompanied by the ejection of one or more neutrons.

$$^{252}_{98}Cf \rightarrow \,^{140}_{54}Xe + \,^{108}_{44}Ru + 4\,^{1}_{0}n$$

For all but the very heaviest isotopes, spontaneous fission is a very slow reaction. Spontaneous fission of ^{238}U, for example, is almost two million times slower than the rate at which this nuclide undergoes α-decay.

NEUTRO N-RICH VERSUS NEUTRON-POOR NUCLIDES

In 1934 Enrico Fermi proposed a theory that explained the three forms of beta decay. He argued that a neutron could decay to form a proton by emitting an electron. A proton, on the other hand, could be transformed into a neutron by two pathways. It can capture an electron or it can emit a positron. Electron emission therefore leads to an increase in the atomic number of the nucleus.

$$^{14}_{6}C \rightarrow\ ^{14}_{7}N +\ ^{0}_{-1}e$$

Both electron capture and positron emission, on the other hand, result in a decrease in the atomic number of the nucleus.

$$^{7}_{4}Be +\ ^{0}_{-1}e \rightarrow\ ^{7}_{3}Li + hv$$

$$^{11}_{6}C \rightarrow\ ^{11}_{5}B +\ ^{0}_{+1}e$$

A plot of the number of neutrons versus the number of protons for all of the stable naturally occurring isotopes is shown in the figure below. Several conclusions can be drawn from this plot.

- The stable nuclides lie in a very narrow band of neutron-to-proton ratios.
- The ratio of neutrons to protons in stable nuclides gradually increases as the number of protons in the nucleus increases.
- Light nuclides, such as ^{12}C, contain about the same number of neutrons and protons. Heavy nuclides, such as ^{238}U, contain up to 1.6 times as many neutrons as protons.
- There are no stable nuclides with atomic numbers larger than 83.

- This narrow band of stable nuclei is surrounded by a sea of instability.
- Nuclei that lie above this line have too many neutrons and are therefore neutron-rich.
- Nuclei that lie below this line don't have enough neutrons and are therefore neutron-poor.

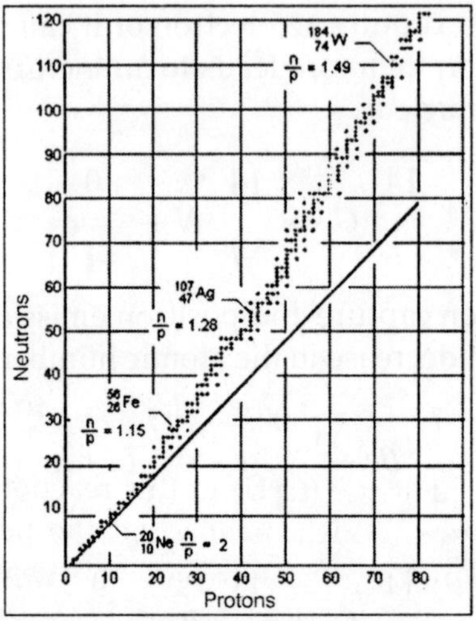

Fig. A Graph of the Number of Neutrons Versus the Number of Protons for all Stable Naturally Occurring Nuclei.

The most likely mode of decay for a neutron-rich nucleus is one that converts a neutron into a proton. Every neutron-rich radioactive isotope with an atomic number smaller 83 decays by electron (β^-) emission. ^{14}C, ^{32}P, and ^{35}S, for example, are all neutron-rich nuclei that decay by the emission of an electron.

$$^{35}_{16}S \rightarrow \ ^{35}_{17}Cl + \ ^{0}_{-1}e \qquad (\beta^- \ emission)$$

Neutron-poor nuclides decay by modes that convert a proton into a neutron. Neutron-poor nuclides with atomic numbers less than 83 tend to decay by either electron capture or positron emission. Many of these nuclides decay by both routes, but positron emission is more often observed in the lighter nuclides, such as ^{22}Na.

$$ {}^{22}_{11}Na \rightarrow {}^{22}_{10}Ne + {}^{0}_{+1}e \quad (\beta^{+} \text{ emission}) $$

Electron capture is more common among heavier nuclides, such as ^{125}I, because the 1s electrons are held closer to the nucleus of an atom as the charge on the nucleus increases.

$$ {}^{125}_{53}I + {}^{0}_{-1}e \rightarrow {}^{125}_{52}Te + h\nu \quad (\text{electron capture}) $$

A third mode of decay is observed in neutron-poor nuclides that have atomic numbers larger than 83. Although it is not obvious at first, -decay increases the ratio of neutrons to protons. Consider what happens during the -decay of ^{238}U, for example.

$$ {}^{238}_{92}U \rightarrow {}^{234}_{90}Th + {}^{4}_{2}He \quad (\alpha \text{ decay}) $$

The parent nuclide (^{238}U) in this reaction has 92 protons and 146 neutrons, which means that the neutron-to-proton ratio is 1.587. The daughter nuclide (^{234}Th) has 90 protons and 144 neutrons, so its neutron-to-proton ratio is 1.600. The daughter nuclide is therefore slightly less likely to be neutron-poor, as shown in the figure below.

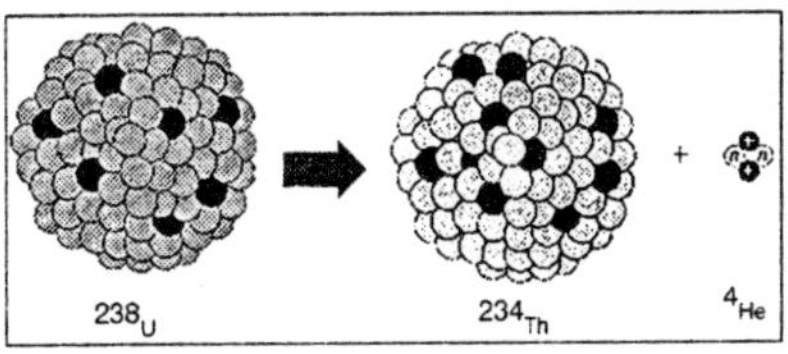

$$ \left(\frac{146 \text{ neutrons}}{92 \text{ protons}} = 1.587 \right) \qquad \left(\frac{144 \text{ neutrons}}{90 \text{ protons}} = 1.600 \right) $$

BINDING ENERGY CALCULATIONS

We should be able to predict the mass of an atom from the masses of the subatomic particles it contains. A helium atom, for example, contains two protons, two neutrons, and two electrons.

Example:

The mass of a helium atom should be 4.0329802 amu.

2(1.0072765) amu = 2.0145530 amu
2(1.0086650) amu = 2.0173300 amu
2(0.0005486) amu = 0.0010972 amu
Total mass = 4.0329802 amu

When the mass of a helium atom is measured, we find that the experimental value is smaller than the predicted mass by 0.0303769 amu.

Predicted mass = 4.0329802 amu
Observed mass = 4.0026033 amu
Mass defect = 0.0303769 amu

The difference between the mass of an atom and the sum of the masses of its protons, neutrons, and electrons is called the mass defect. The mass defect of an atom reflects the stability of the nucleus. It is equal to the energy released when the nucleus is formed from its protons and neutrons. The mass defect is therefore also known as the binding energy of the nucleus.

The binding energy serves the same function for nuclear reactions as H for a chemical reaction. It measures the difference between the stability of the products of the reaction and the starting materials. The larger the binding energy, the more stable the nucleus. The binding energy can also be viewed as the amount of energy it would take to rip the nucleus apart to form isolated neutrons and protons. It is therefore literally the energy that binds together the neutrons and protons in the nucleus. The binding energy of a nuclide can be calculated from its mass defect with Einstein's equation that relates mass and energy.

$$E = mc^2$$

Example: We found the mass defect of He to be 0.0303769 amu. To obtain the binding energy in units of joules, we must convert the mass defect from atomic mass units to kilograms.

$$0.030769\,amu\,x\frac{1.6605655\,x10^{-24}\,g}{1\,amu}\,x\frac{1kg}{1000g}=5.04428x10^{-29}\,kg$$

Multiplying the mass defect in kilograms by the square of the speed of light in units of meters per second gives a binding energy for a single helium atom of 4.53358×10^{-12} joules.

$$E = (5.04428x10^{-29}\,kg)(2.9979246x10^{8}\,m/s)^{2} = 4.53358x10^{-12}\,J$$

Multiplying the result of this calculation by the number of atoms in a mole gives a binding energy for helium of 2.730 x 10^{12} joules per mole, or 2.730 billion kilojoules per mole.

$$\frac{4.5358x10^{-12}\,J}{1\,atom} x \frac{6.022x10^{23}\,atoms}{1\,mol} = 2.730x10^{12}\,Jimol$$

This calculation helps us understand the fascination of nuclear reactions. The energy released when natural gas is burned is about 800 kJ/mol. The synthesis of a mole of helium releases 3.4 million times as much energy. Since most nuclear reactions are carried out on very small samples of material, the mole is not a reasonable basis of measurement. Binding energies are usually expressed in units of electron volts (eV) or million electron volts (MeV) per atom.

Example: The binding energy of helium is 28.3 × 10^6 eV/atom or 28.3 MeV/atom.

$$\frac{4.5358x10^{-12}\,J}{1\,atom} x \frac{1\,ev}{1.6021892x10^{-19}\,J} = 28.30x10^{6}\,evlatom$$

Calculations of the binding energy can be simplified by using the following conversion factor between the mass defect in atomic mass units and the binding energy in million electron volts.

1 amu = 931.5016 MeV

Binding energies gradually increase with atomic number, although they tend to level off near the end of the periodic table. A more useful quantity is obtained by dividing the binding energy for a nuclide by the total number of protons and neutrons it contains. This quantity is known as the binding energy per nucleon. The binding energy per nucleon ranges from about 7.5 to 8.8 MeV for most nuclei, as shown in the figure below.

It reaches a maximum, however, at an atomic mass of about 60 amu. The largest binding energy per nucleon is observed for ^{56}Fe, which is the most stable nuclide in the periodic table.

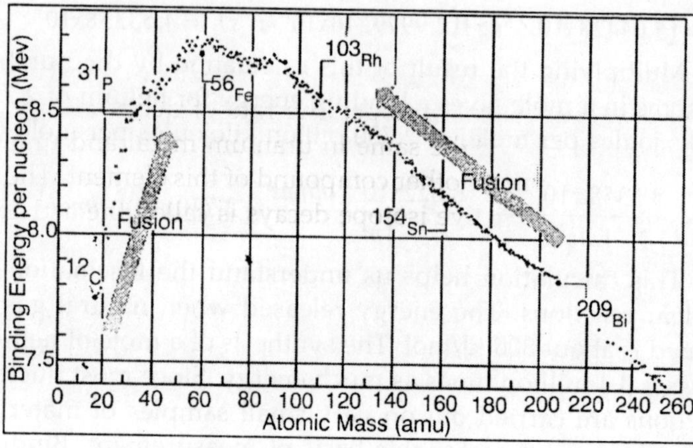

Fig. Graph of Binding Energy

The graph of binding energy per nucleon versus atomic mass explains why energy is released when relatively small nuclei combine to form larger nuclei in fusion reactions.

$$^{12}_{6}C + ^{12}_{6}C \rightarrow ^{24}_{12}Mg \quad (\textit{fusion})$$

It also explains why energy is released when relatively heavy nuclei split apart in fission (literally, "to split or cleave") reactions.

$$^{235}_{92}U \rightarrow ^{139}_{56}Ba + ^{94}_{36}Kr + 2\,^{1}_{0}n \quad (\textit{fission})$$

There are a number of small irregularities in the binding energy curve at the low end of the mass spectrum, as shown in the figure below. The ^{4}He nucleus, for example, is much more stable than its nearest neighbors. The unusual stability of the ^{4}He nucleus explains why -particle decay is usually much faster than the spontaneous fission of a nuclide into two large fragments.

THE KINETICS OF RADIOACTIVE DECAY

Radioactive nuclei decay by first-order kinetics. The rate of radioactive decay is therefore the product of a rate constant

(k) times the number of atoms of the isotope in the sample (N).

$$\text{Rate} = kN$$

The rate of radioactive decay doesn't depend on the chemical state of the isotope. The rate of decay of ^{238}U, for example, is exactly the same in uranium metal and uranium hexafluoride, or any other compound of this element. The rate at which a radioactive isotope decays is called the activity of the isotope.

The most common unit of activity is the curie (Ci), which was originally defined as the number of disintegrations per second in 1 gram of ^{226}Ra. The curie is now defined as the amount of radioactive isotope necessary to achieve an activity of 3.700×10^{10} disintegrations per second.

The relative rates at which radioactive nuclei decay can be expressed in terms of either the rate constants for the decay or the half-lives of the nuclei. We can conclude that ^{14}C decays more rapidly than ^{238}U, for example, by noting that the rate constant for the decay of ^{14}C is much larger than that for ^{238}U.

^{14}C:		$k = 1.210 \times 10^{-4}\ y^{-1}$
^{238}U:		$k = 1.54 \times 10^{-10}\ y^{-1}$

We can reach the same conclusion by noting that the half-life for the decay of ^{14}C is much shorter than that for ^{235}U.

^{14}C:		$t_{1/2} = 5730\ y$
^{238}U:		$t_{1/2} = 4.51 \times 10^{9}\ y$

The half-life for the decay of a radioactive nuclide is the length of time it takes for exactly half of the nuclei in the sample to decay. In our discussion of the kinetics of chemical reactions, we concluded that the half-life of a first-order process is inversely proportional to the rate constant for this process.

$$t_{1/2} = \frac{\text{In } 2}{k} = \frac{0.693}{k}$$

The half-life of a nuclide can be used to estimate the amount of a radioactive isotope left after a given number of half-lives.

DATING BY RADIOACTIVE DECAY

The earth is constantly `bombarded by cosmic rays emitted by the sun. The total energy received in the form of cosmic rays is small no more than the energy received by the planet from starlight. But the energy of a single cosmic ray is very large, on the order of several billion electron volts (0?200 million kJ/mol). These highly energetic rays react with atoms in the atmosphere to produce neutrons that then react with nitrogen atoms in the atmosphere to produce ^{14}C.

$$^{14}_{7}N + ^{1}_{0}n \rightarrow ^{14}_{6}C + ^{1}_{1}H$$

The ^{14}C formed in this reaction is a neutron-rich nuclide that decays by electron emission with a half-life of 5730 years.

$$^{14}_{6}C \rightarrow ^{14}_{7}N + ^{0}_{-1}e$$

Just after World War II, Willard F. Libby proposed a way to use these reactions to estimate the age of carbon-containing substances. The ^{14}C dating technique for which Libby received the Nobel prize was based on the following assumptions.

- ^{14}C is produced in the atmosphere at a more or less constant rate.
- Carbon atoms circulate between the atmosphere, the oceans, and living organisms at a rate very much faster than they decay. As a result, there is a constant concentration of ^{14}C in all living things.
- After death, organisms no longer pick up ^{14}C.

Thus, by comparing the activity of a sample with the activity of living tissue we can estimate how long it has been since the organism died.

The natural abundance of ^{14}C is about 1 part in 10^{12} and the average activity of living tissue is 15.3 disintegrations per minute per gram of carbon. Samples used for ^{14}C dating can include charcoal, wood, cloth, paper, sea shells, limestone, flesh, hair, soil, peat, and bone. Since most iron samples also contain carbon, it is possible to estimate the time since iron was last fired by analyzing for ^{14}C.

We now know that one of Libby's assumptions is questionable: The amount of ^{14}C in the atmosphere hasn't been constant with time. Because of changes in solar activity and the earth's magnetic field, it has varied by as much as 5 per cent. More recently, contamination from the burning of fossil fuels and the testing of nuclear weapons has caused significant changes in the amount of radioactive carbon in the atmosphere. Radiocarbon dates are therefore reported in years before the present era (B.P.). By convention, the present era is assumed to begin in 1950, when ^{14}C dating was introduced.

Studies of bristlecone pines allow us to correct for changes in the abundance of ^{14}C with time. These remarkable trees, which grow in the White Mountains of California, can live for up to five thousand years. By studying the ^{14}C activity of samples taken from the annual growth rings in these trees, researchers have developed a calibration curve for ^{14}C dates from the present back to 5145 B.C.

After roughly 45,000 years (eight half-lives), a sample retains only 0.4 per cent of the ^{14}C activity of living tissue. At that point it becomes too old to date by radiocarbon techniques. Other radioactive isotopes can be used to date rocks, soils, or archaeological objects that are much older. Potassium-argon dating, for example, has been used to date samples up to 4.3 billion years old. Naturally occurring potassium contains 0.0118 per cent by weight of the radioactive ^{40}K isotope.

This isotope decays to ^{40}Ar with a half-life of 1.3 billion years. The ^{40}Ar produced after a rock crystallizes is trapped in the crystal lattice. It can be released, however, when the rock is melted at temperatures up to 2000C. By measuring the amount of ^{40}Ar released when the rock is melted and comparing it with the amount of potassium in the sample, the time since the rock crystallized can be determined.

Chapter 4

Composite Structure of an Atom

BOHR'S MODEL

The Danish physicist Niels Henrik David Bohr (1885-1962) has created the theory of a structure of atom, which was deprived of the basic deficiencies of Rutherford's model of atom.

Thus he had to change some habitual representations of classical physics. Bohr has formulated three postulates (sometimes the first and the second postulates are combined, then Bohr's theory of atom is reduced to two postulates):

- The electrons in atom rotate around the kern on particular stationary orbits only. The motions on these orbits correspond to stationary states of atom, which do not vary in time without exterior actions. Rotating on stationary orbits, the electrons do not radiate energy.
- Allowed, discrete values of energy of an electron correspond to stationary states of atom.

This postulate was called the quantization rule of orbits. It affirms, that when the electron is going on a stationary orbit, it should have discrete values of a moment of motion:

$$m_0 \upsilon r = n\frac{h}{2\pi}, n = 1,2,3,...$$

m_0 & nbsp of electronic mass; υ -velocity of atom;

r -radius of an orbit h -Planck's constant;

$$h = 6,62 \times 10^{-34}$$

- The radiation or uptake of energy by atom occurs only at transition of an electron from one stationary orbit to another. Thus quantum of light energy (photon) is radiated or is immersed. The energy of a quantum is equal to a difference of energies of an electron on the relevant stationary orbits: frequency of a light, energy of an electron on m and n orbits.

The laws of classical physics say, that the transition of bodies from one state to another occurs continuously at continuous radiation or uptake of energy. The classical laws have appeared inapplicable to the phenomena, which are described by Bohr.

Parameters of Atom

The radiuses of stationary orbits and energies of electrons are determined by dependence of energy of atom on radius of an electronic orbit. The elementary system is the hydrogen-like atom. The hydrogen-like atom is the one, in which one electron rotates on a circular orbit around of a kern, which has a charge *Ze* (Z - serial number of a device in periodic system).

The energy of an electron in atom consists of a kinetic energy of a motion on an orbit and potential energy in an electric field of a kern (electronic mass, travelling speed of an electron on an orbit, radius of an orbit, permittivity of vacuum of empty space):

$$E = \frac{m_0 \upsilon^2}{2} + \frac{-Ze^2}{2} - \frac{Ze^2}{4\pi\varepsilon_0 r}$$

m_0 & nbsp -electronic mass; υ -velocity of atom;

ε_0-permittivity of vacuum of empty space'

r -radius of an orbit;

The kinetic energy is determined with the requirement, that a centripetal acceleration is created by force of a Coulomb attraction of charges. The kinetic energy of an electron is inversely proportional to radius of an orbit. A total energy:

$$E = \frac{1}{2}\frac{Ze^2}{4\pi\varepsilon_0 r} - \frac{Ze^2}{4\pi\varepsilon_0 r} = -\frac{1}{2}\frac{Ze^2}{4\pi\varepsilon_0 r}$$

The total energy of an electron varies with the change of a radius of an orbit the same as potential energy does. Hence, the more than radius of an orbit of an electron, which rotates around the kern, the larger is the total energy of atom.

Value $r_1 = 0{,}528 \times 10^{-10}$ m is called the first Bohr's radius. For the subsequent values: $r_n = n^2 r_1$

n – main quantum number. It specifies an energy level, on which the atom is. The lowest state of atom is inconvertible. All atoms are in the lowest state at standard conditions. The atom can be exited, if the energy transferred to it. Thus it will be transferred to one of higher energy states.

The distance from the lowest energy level $n = 1$ up to the proximate level with higher energy $n = 2$ makes 10,1 $eV = 16{,}2 \times 10^{-19}$ J for the atom of hydrogen. This energy is the least bundle of energy, which the atom of hydrogen can immerse, when it is in a normal state.

Deficiencies of the Theory

Bohr's theory explained the nature of a spectral series of hydrogen. It has allowed to spot theoretically the sizes of atom and energy levels of electrons for atom of hydrogen. Deficiency of Bohr's theory is the inconsistency of its build-up. Bohr's postulates have quantum character, but the stationary electronic orbits and allowed levels of energy of atom are determined by methods of a classical mechanics and electrodynamics. Therefore Bohr's theory has appeared usable to calculations only of monovalent atoms.

Bohr's theory does not give an opportunity to calculate intensity of spectral lines, which is interlinked to probabilities of electronic transitions. The introduction of quantum rules for definition of orbits and levels of electrons was not logically justified by Bohr.

The further development of the Bohr's theory was made by German physicist Arnold Johannes Wilhelm Sommerfeld (1868-1951). He has changed circular orbits on elliptic. It has enabled to apply Bohr's theory to multi electronic atoms. However the creation of a fundamental theory of atom is bound to the creation of a quantum mechanics.

RUTHERFORD'S ATOMIC MODEL

The perception of an atom as the least indivisible particle of substance was the foundation of the cinematic gas theory. It have explained and predicted a series of the experimental phenomena. However, at the end of XIX century new experimental data proved the composite structure of an atom. The experimentally found laws of radiation and uptake of energy by atoms (spectral series, radiation of an ideal black body, photoeffect) have not received an explanation within the framework of a classical electrodynamics. Only new ideas, based on the quantum theory, helped explaining these phenomenon.

Structure of Atom

British physicist Ernest Rutherford (1871-1937) has created a planetary model of atom exploring a dispersion of a-particles transiting through a thin metal foil. According to this model an atom looks like a tiny planetary system, in which the forces of an electrical attraction operate. The centre of an atom is a positively charged kern. Almost all mass of an atom is focused there. Negatively charged eletrons rotate around the kern.

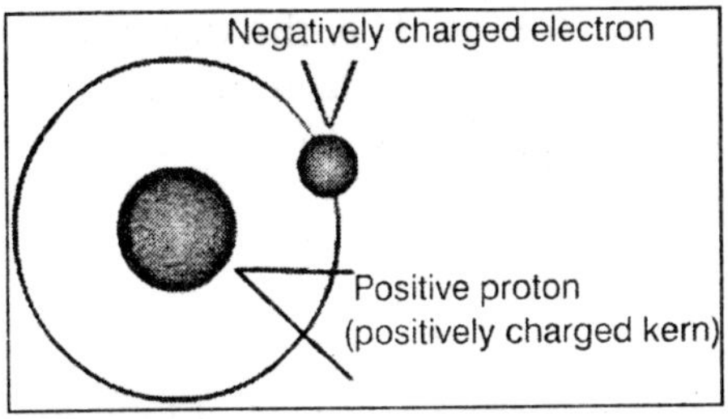

Fig. An Atom of Hydrogen

An atom of hydrogen has the simplest structure. Positively charged kern of an atom of hydrogen was called a proton. The only electron rotates around it. It was determined later, that every atom's kern consists of protons and neutrons. The neutron's mass is almost equal to protons, but has not electric charge. The mass of an electron is about 1/1840 of a proton's mass and does not influence an atomic weight of substance.

The number of positive and negative charges of atom determines the position of a substance in a periodic system. The positive charge of an atom's kern is equal to a serial number of a substance in the table.

The number of neutrons in an atom's kern is equal to a residual between a rounded atomic weight and a number of positive protons in a kern (serial number of substance). The electrical neutrality rule of atom says that the number of electrons rotating around the kern is equal to a number of positive protons contained in an atom.

Parameters of Atom

According to the Rutherford's experimental results the radius of a trajectory of an electron in a hydrogen atom is $0{,}53 \times 10^{-10}$ m, the diameter of an atom of hydrogen is equal to $2{,}6 \times 10^{-15}$ m. Volume of electrons and kern is approximately 1/10 000 of an atom's volume. The substance of a kern has huge density of 200 million ton in 1 cm^3. It is all because the most of an atom's mass is concentrated in its kern.

The mass of a positive proton is approximately equal to the mass of a neutron and is $1,\ 67 \times 10^{-27}$. The mass of an electron is $9,\ 11 \times 10^{-31}$ kg. The classic radius of an electron, r_0, is equal to $2,\ 8 \times 10^{-17}$ m. It is the distance, on which the electrostatic energy of an electron intercourse with a point charge of the same quantity is equal to natural energy of an electron:

$$\frac{e^2}{r_0} = m_0c^2$$

e – elementary charge; c – speed of light in vacuum; m_0 – electronic mass

Deficiencies of the Theory

The Rutherford model of atom was well correlated with the results of a dispersion of an a-particle by atoms of substance, but it has not explained neither the process of radiation of atoms, nor legitimacies in spectrums of radiation. According to the laws of an electrodynamics rotating around the kern the electron should radiate electromagnetic waves. As a result of radiation the natural energy of an electron should be diminished, thus the trajectory it will be figured by a spiral, and during the order 10^{-8} about an electron should fall on a kern. Such deduction obtained on the basis of representations of classical physics about radiation, contradicted known stability of atoms and character of nuclear spectrums of radiation.

MULTIELECTRONIC ATOMS

Structure of Atom

Just as in the atom of hydrogen, in more composite atoms electrons can rotate around of a kern on allowed orbits only. The orbits in composite atoms cluster in system of shells. Each shell contains a particular number of orbits, an electron can be on any og them. A *K*-shell is the closest to a kern and contains two orbits. The second is a *L*-shell has eight orbits. The third is a *M*-shell has eight orbits, the fourth *N*-shell has 18 orbits etc.

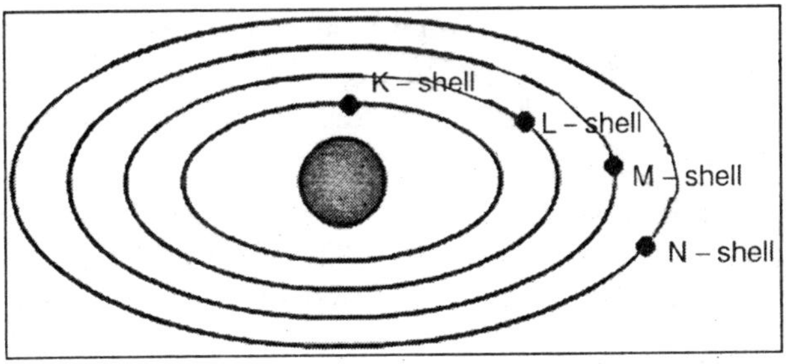

Fig. Structure of Atom

Parameters of Atom

When the electron travels from an orbit of greater radius to an orbit of smaller radius, the energy is oozed. But the transition from outer shells on interior is possible only when there are free orbits on the leter. The outer-shell electrons are less bound to a kern than interior ones: they are on greater distance from a kern, and the electrons, which are on inner shells, screen the nuclear charge.

Therefore just a small amount of energy is sufficient for excitation of outer-shell electrons. The quantums of energy radiated at transition of such electrons in a nonexcited state have a quantity of about several electron-volt. Greater energy is necessary to detach inner-shell electrons from the atom.

The greater the nuclear charge the greater energy is needed. Therefore, when there is a transition of electrons from outer shells to a *K*-shell in heavy atoms, the emission of quantums with energy of hundreds and even thousand electron-volt is possible. Such quantums of high energy are called Roentgen rays.

X-ray Spectrums of Atoms

The X-rays results from losses by atom of an inner-shell electron: The transition of electrons is accompanied by an emission of quantums of high energy with a small wave length, which are termed as Roentgen rays.

The X-rays emerges when atom losses inner-shell electron: The transition of electrons is accompanied by an emission of quantums of high energy with a small wave length called Roentgen rays. The atom can lose inner-shell electrons as a result of an electron bombardment with electrons speeded up to energy of several thousand electron-volt.

The wave length of the X-rays is comparable to interatomic distance in solid bodies. The examination of the X-ray diffraction on crystal lattices of solid bodies enables, knowing a wave length of Roentgen rays, to determine interatomic distances in them.

The examination of Roentgen rays also allows to determine nuclear charge, because it depends on a prime relation of an electron-binding energy in atom, which is expressed through energy of the X-rays. This relation is called Moseley's law:

$$v = A\,(Z - B)$$

v – frequency of a line in a X-ray

Characteristic spectrum of a device;

A and B-constants; Z -serial number of a device in periodic system

PROPERTIES OF LIGHT

In one phenomena, which are bound to distribution of light, its undular properties (interference, diffraction) are shown. In other phenomena its quantum properties (photoeffect, radiation and uptake of a light, process of interaction of a light with substance) are shown. The light has simultaneously undular and quantum properties, which are differently shown in the different phenomena. However, when the wave length of a light is diminished, its undular properties are attenuated and the quantum properties are shown.

Compton's Effect

The experience of Arthur Holly Compton (1892-1962) on a dispersion of Roentgen rays by substances, which consist of atoms of light devices, have shown, that the dispelled Roentgen rays have a major wave length l¢, than impinging l. Thus the residual Dl = l¢ - l depends on properties of scattering substance and wave length of an impinging light:

$$\Delta\lambda=\lambda' - \lambda = 2\lambda_K \sin^2 \frac{\theta}{2}$$

The quantity λ_K is a stationary value for all substances also is termed as Compton's wave length, $\lambda_K = 2{,}43 \times 10^{12}$ m. This phenomenon, which is termed as Compton's recoil, is impossible to explain by a wave theory of light. In quantum theory the magnification of wave length of radiation at a dispersion is explained by interaction of photons of impinging radiation with electrons of atoms and molecules.

The part of energy of an impinging photon, which is transmitted to electrons of scattering substance, depends on a scattering angle of photon. At $\theta = \pi$, that is provided that after a dispersion a photon will fly in the party, which is opposite to a tentative direction, the losses by a photon of energy will be maximal, that will give in maximal magnification of a wave length of radiation.

Light Pressure

Some properties, for example light pressure, can be

surveyed simultaneously in the electromagnetic theory and in the quantum theory.

According to the electromagnetic theory of light, the pressure p, which renders a luminous flux with power W, impinging normally on a simple surface, is equal:

$$p = \frac{W}{c}(1+R)$$

For an ideal black body (R = 0):

$$p = \frac{W}{c}$$

For an absolute reflecting body (R = 1):

$$p = \frac{2W}{c}$$

If N of quantums of a luminous flux with frequency n, which bears(carries) energy $W = N \cdot h$n, impinges on a simple surface for 1 second, then the blanket impulse at a reflectivity R will look like:

$$p = \frac{N \cdot hv}{c}(1+R) = \frac{W}{c}(1+R)$$

Thus, electromagnetic theory and quantum theory give identical results. However it is essential different representations will be utillized in the electromagnetic and quantum theories of light. The concept of an electromagnetic wave of application to a light guesses existence of a wave in particular field(area) of space, which is capable to contain even some lengths of waves of light.

The light quantums are considered as a particle, which are localized in very small volume of space. The representation about a light wave guesses a continuous distribution of energy in a spatial wave, the representation about quantums guesses localization of energy in a small part of space. Both these incompatible representations reflect unity of a wave-corpuscle nature of light, which is shown at distribution of light and its interaction with substance.

PROPERTIES OF SUBSTANCE

De Broglie's Theory

The French physicist Louis Victor duc de Broglie has applied wave-corpuscle treatment and to particles of substance: as the light detects corpuscular properties, also particles of substance should detect undular properties. De Broglie has utillized already wave-corpuscle properties, accepted for a light, and their quantum rule and has expressed a wave length through the performances of a particle. For a light the corpuscular point of view determines energy and impulse accordingly by expressions $E = mc^2$ and $p = mc$.

The wave length of a particle of substance is determined under the formula:

$$\lambda = \frac{h}{p} = \frac{h}{m\upsilon}$$

At adding in a rule of quantization of a moment of momentum of an electron the impulse of an electron, will turn out a relation $n\lambda = 2\pi r$. From it follows, that on length of a stationary orbit of an electron the integer of lengths of waves should be stacked. Thus, the requirement of existence of allowed orbits, which was injected by Bohr frequency relation, acquires legible physical sense: the allowed orbits are orbits supposing formation on them of standing waves of an electron. For a rectilinear motion of particles de Broglie has offered expression, which is made by analogy with the equation for a spread flat light wave in the complex shape:

$$\psi = \psi_0 e^{i\left(\omega t - \frac{x}{\lambda}\right)}$$

De Broglie's Waves

The specified waves have received a title of associated waves. Thus, the associated waves are traveling waves for freely propellent electrons or standing waves for electrons, which are bound in atoms. De Broglie's waves are not waves of propellent substance, they have not analog in classical physics. De Broglie's guess about existence of undular properties of particles carries universal character: undular properties should have the electron, positive proton, neutron, atom, molecule, any moving object.

However objects, which have major masses and move with usual velocities, will have a so small wave length To in comparison with the sizes of objects, that the phenomena of an interference and diffraction for them can completely be neglected. The undular properties of a light are clearly shown in cases, when the wave length has compared to the sizes of bodies, with which the light interreacts.

The lengths of waves of electrons in usual requirements have the order of the nuclear sizes, hence, the effects are characteristic for them which are observed usually for Roentgen rays. For such small lengths of waves it is possible to observe a diffraction on nuclear crystal lattices.

EXAMINATION OF PROPERTIES

The existence of undular properties of substance was revealed at observation of patterns of a diffraction of beams of particles. Legible patterns of an interference and the diffractions give bundles of electrons, which have almost identical impulses. These patterns are very similar to optical patterns of an interference and diffraction.

At change of quantity of a momentum of electrons and measuring of a wave length, it is possible to receive a relation between these quantities. It has appeared equal to a known relation

$$\lambda = \frac{h}{p}$$

This de Broglie's relation is valid for any kind of particles. The undular properties of particles were detected at observation of a pattern of the diffraction of the bundle of electrons on chips. At measuring a scattering angle it is possible to spot a wave length of electrons. Thus, at known quantity of potential, which was utilised for acceleration of a bundle of electrons, their velocity is possible to calculate, and then to receive independent value of a wave length of an electron. The value of a wave length of an electron, which was spotted in such a way, completely coincides with quantity of a wave length, which is found from a diffraction pattern.

The undular properties of electrons also were detected in experience with a cathode rays at major velocity, which was passed through a thin metal foil on a photographic plate. After development of a photoplate on it the diffraction pattern from rings of different intensity was visible. However this diffraction pattern was stipulated not by Roentgen rays, which could arise at a concussion of high-velocity electrons with metal. It confirms presence of a stream of charged particles.

MECHANICS

Quantum Mechanics is the study of matter and energy on the subatomic scale. It was came into being because classical physics could not explain certain experimental results, such as why the photoelectric effect occured only when the light shone is above a certain frequency. Quantum Mechanics and Relativity became the foundation of modern physics.

Main Points

Principle of Quantatization

The fundamental principle of Quantum Mechanics, and its namesake, is that energy came in discrete particles called "quanta". This idea, the Quantatization of Energy, was put forth by Planck in his Quantum Hypothesis(1900), an explanation of blackbody radiation. It was a departure from Classical Physics, which asserted that energy could be infinately small.

Photoelectric Effect

This quantatization of energy approach was used by Einstein in his 1905 explanation of the photoelectric effect, the emission of electrons by a metal when light is shone on it. No electrons are emitted when the light is below a threshold frequency, no matter how intense the light.

Einstein postulated that light, while exibiting wave properties as shone by the Double-slit Experiment, also came in the from of particles called "photons", discrete packets of energy. The energy of a photon is proportional to the frequency of the light ($E = hv$); thus, light with higher frequency had more energy and could excite the electrons, while light whose frequency were below the threshold could not, no matter how many photons there were.

Bohr's Model and Spectral Lines

Bohr also used the Quantatization of energy in his 1913 model of the atom. He stated that electrons revolved around the nucleus much like the planets revolve around the sun. They could be found in predefined, stationary orbits, radiating no energy as they revolve. Only when they gain a quatum of energy equivalent to the difference between the energy of the orbits do they become excited and enter a higher energy orbit. This is consistent with Einstein's explanation of the photoelectric effect, and explains why spectral lines only occur at certain places.

Wave-Particle Duality of Matter

In 1923 De Broglie expanded the Principle of Wave-particle Duality of Light put forth by Einstein to include all matter, called Principle of Wave-particle Duality of Matter. He asserted that all matter, not just light, exibit both wave-like and particle-like behaviour,

Quantum Matrix and Wave Mechanics

By now there was enough theories to put together a quantum picture of the atom. However, it was very difficult to sort through all the math. Heisenherg and Shroedinger simultaneously and independently worked to put forth a complete quantum mechanics. Heisenberg employed use of matrices, and his model was called the matrix mechanics.

His approach emphasized the quantum—discrete properties while Schroedinger concentrated on the wave properties. He tried to find a function that would encompass all informatin about a particle or system at any given time. This is called the wave function. While debated ensued about which was correct, Schoedinger proved shortly after that the matrix mechanics and the wave function were mathematically equivalent.

Uncertainty Principle

Heisenberg then came up with the Uncertainty Principle, which stated that the position and velocity(momentum, which is mass*velocity) cannot be known precisely simultaneouely. The more accurately one is known, the less accurately the other can be measured. The product of the uncertainty in position, and the uncertainty in momentum, is always greater than or equal to $h/(2pi)$. While this is neglegible on the macroscopic level, subatomically, it causes great uncertainty. As a result, the world exists in statistical probabilities.

For example, a electron cannot be found precisely at one location. At any given chance, it has a certain probability of being found anywhere in the universe, with higher probabilities in regions around the atomic nucleus and less likelihood elsewhere. This can be calculated with the wave function.

Quantum Atomic Model

The quantum model of the atom calls for arbitrarily defined orbitals. An orbital is a region around the nucleus where an electron has 90 per cent chance of being found. Electrons have spin, a term used for the intrinsic angular momenta subatomic particles have. They have either a spin of +1/2 or –1/2. Two electrons in the same orbital of the same energy level must have opposite spin (Pauli Exclsion Principle).

Schroedinger's Cat

Since the wave function describes the world in terms of statistical probabilities and not certainties, a cat placed in an isolated system with equal chances of being found alive and dead at anytime *is* equally dead and alive at the same time.

When an observer is introduced to the isolated system, the wave function collapses and the cat is either dead or alive.

PRINCIPLES OF QUANTUM MECHANICS

UNCERTAINTY PRINCIPLE

One of the most important concepts of quantum mechanics is the uncertainty principle, which states that there is always some uncertainty when trying to measure the posistion and momentum of a particle. More specifically, the product of the uncertainty in position, D x, and the uncertainty in momentum, D*m*u, is always greater than or equal to h over 2p, where h is planck's constant.

When trying to determine the position of an electron, for example, we cannot know *exactly* where it is and where it's going at the same time. The more accurately one is measured, the less accurately the other is known.

To we measure the position of the electron, for example, we can only obtain the information through the collision of the electron with another subatomic particle, such as a photon. After the collision, the momentum of the photon changes—it "bounces" away, and from this change in momentum the position of the electron is calculated (this is how we see-photons bouncing off objects make possible vision).

However, the photon also gives the electron a change in momentum, so that now the position and velocity of the electron has changed from what it was. Thus, we can't know precisely the position and velocity simultaneously.

ELECTRON IN THE BOHR ATOM

In the Bohr model, electrons revolved around the nucleus in orbits. According to classical physics, they should radiate energy as they move and ultimately fall to the centre of the atom.

Yet this is not so. The Quantum explanation is that electrons can only exist in predefined, discrete energy levels, and nowhere in between. The lowest energy level is not at the centre of the atom, and electrons may go no closer than the lowest level.

This explains why they don't collide into the nucleus. To move to another energy level, they must gain or lose *exactly* the energy difference between the levels. This explains spectral lines. Another explanation is that, if electrons crashed into the nucleus, their velovity would be zero and the uncertainty would also be zero. This would be a direct violation of the uncertainty principle, a fundamental property of nature.

SPIN AND QUANTUM NUMBERS

It was discovered experimentally that subatomic particles have intrinsic angular momentum. That is, that angular momentum is not given by some interaction, but a property of the particles. This is a quantum property that classical physics has no explanation for. The closest thing on the macroscopic level is rotation (since angular momentum leads to rotating), so this property is called "spin". Subatomic particles, however, don't really spin; they just have the angular momenta. This property is quantatized since the particles may only have certain discrete values of spin and not any value in between. Electrons have spins of either +1/2 or –1/2.

In the quantum model, electrons in the atom are given four quantum numbers to distinguish them. The first three distinguish the energy level, the shape of the orbital, and in which orbital the electron is located (has 90 per cent chance of being found), and the last one gives the spin. The Pauli Exclusion Principle states that no 2 electrons in the atom can have the same 4 quantum numbers. If they occupy the same energy level and orbital, then they must have opposite spin.

NUCLEUS

The nucleus is a compact and dense region in the centre of the atom that makes up for most of its mass. It is made up of a group of subatomic particles known collectively as nucleons. The two states of nucleons are protons and neutrons.

Although the electron, and not the nucleus, is responsable for chemical reactions, the nucleus still holds much information about the atom. Since nucleons are about 2000 times heavier than electrons, the mass of the nucleus is approximately equal to the mass of the atom. The number of protons in the nucleus determines the type of atom, and nuclear charge is the property by which the elements are arranged in the periodic table.

NUCLEAR ENERGY

A tremendous amount of energy is used to hold the nucleus together, since the repulsion between the positively charged protons is great. The bonding forces that hold the nucleus together are collectively known as the nucleur forces. When heavy element nuclei are split into smaller nuclei, or light element nuclei combined to make one nucleus, tremendous energies are released. The former is known fission, and the latter, fusion, and they are the driving forces behind stars and nuclear weopons.

From Dalton's theory, it was thought that atoms were indivisible particles that made up everything in the world. When Rutherford made his gold foil experiment, however, he found that when high-velocity positive alpha particles bombarded a thing metal foil, while most passed straight through, some were reflected back or deflected. This led him to postulate that most of the atom was empty space, with most of the mass concentrated in one dense region in the centre of the arom.

Thus, most of the positive alpha particles passed through the empty space, but some hit the dense core and were reflected back by electric repulsion. He called this the nucleus of the atom. Subsequently, improvements were made on the Rutherford model. The nucleu was theorized to contain both positive and neutral particles.

APPLICATION

The interactions in the nucleus, the strong and the weak forces, are the strongest forces in nature. Processes such as fission and fusion release much of this energy, and make possible new energy sources. In fact, fusion is the process by which stars generate light and heat, and thus the source from which all life on earth is derived. Other processes, such as radiation, have medical applications.

STRUCTURE OF NUCLEUS

The Dalton atomic model suspected no nucleus. It assumed that atoms were the fundamental building blocks of nature and no particles exist that are smaller than atoms. However, Rutherford's Gold-foil experiment led him to suspect that atoms had nuclei where most of its mass was located, and that the nucleus was positive. Bombarding Nitrogen with alpha particles, he noticed that the disintegration of the nucleus produced a positive particle equal to the mass of a hydrogen atom. He called this particle the proton and postulated that they made up the nucleus. But a lot of positive charge concentrated in a small region would cause a lot of repulsion forces. So when Chadwick discovered the neutron in 1932, Heisenberg created the proton-neutron model of the nucleus. It also successfully explains isotopes.

THE PROTON

The proton is the nucleon that has a positive unit chage. Its mass is one atomic mass unit (amu, or 1.67×10^{-27} kg, 1,836 times the mass of an electron). It is present in the nucleus of every element. Protons are responsible for the nuclear charge, the positive charge of the nucleus that deflected Rutherford's alpha particles. This charge also corresponds to the atomic number, which determines what kind of element the atom belongs to, and where on the periodic table the element belongs.

THE NEUTRON

THe neutron is an electrically neutral particle slightly heavier than the proton. They are present in numbers greater than or equal to the number of protons in every element except Hydrogen (whose nucleus is a single proton), and variations in the number of neutrons of elements accound for isotopes.

The neutron has a maganetic moment, which suggests that its internal structure is made up of electrically charged components whose net charge is zero. The neutron is actually the fusion of a proton and an electron (since in beta decay a neutron is turned into a proton and an electron) and a neutrino.

ISOTOPES

Isotopes are atoms of the same element that have differenet mass. This is explained by the proton-neutron model of the nucleus. The number of protons account for which element it is and thus how it behaves, but the number of neutrons there are determines which isotope it is, and there can be more neutrons than there are protons.

For example, elemental Hydrogen has one proton (H-1) and a mass of one amu. But an isotope of Hydrogen, deuterium, has a proton and a neutron, and a mass of two amu. As another example, Carbon-12, Carbon-13 and Carbon-14 all have 6 protons (this determines that it is Carbon atom) but have 6, 7 and 8 neutrons, respectively.

Sometimes there are too many neutrons in the nucleus. When this happens, the neutron goes under beta decay, a type of radioactive decay. The neutron splits into a proton, an electron (beta particle) and a neutrino. The proton stays in the nucleus, but the beta particle and the neutrino(an electrically neutral, light or massless particle) are emitted, and energy is released, carried by the electron (when the electron doesn't carry all the energy, the rest is carried by the neutrino).

CLASSIFICATION

Nucleons are generally knoen as a baryon. Baryons are elementary particles consisting of three quarks and no antiquarks. They obey the Fermi-Dirac statistics, which encompass all elementary particles that obey the Pauli Exclusion Principle. The proton is the lightest baryon and is made of two up-quarks and one down-quark. The neutron is made of two down-quarks and one up-quark.

Nucleon's antiparticles are the antiproton and the antineutron, which have the same mass as the corresponding nucleons but opposite electric charge and magnetic moment.

TYPES OF RADIOACTIVITY

The study of radioactive material started with Becqueral in 1896, when he discovered that some of his films became unuseable, as if they had been exposed to light, even though they were kept in a sealed opaque container. Further investigation revealed a substance that gave a green glow in the dark. This was uranium, and Becqueral discovered the X-ray.

Marie Curie then studied radioactive material, discovering several radioactive substances. This research led to a better understanding of radiation, and peopls started bombarding elements with particles emitted from radiation. Rutherford's experimentations, for example, led him to his atomic model. These bombardments also led to the discovery of many particles, such as the proton. The nature of radiation was also better understood as the identity of the radiated particles came to be known, and theories of radioactive decay were devised.

These experiments did not reveal a complete picture of radiation, until Fermi's systematic bombardment of all elements. The results led him to envision a chain reaction, where the bombardment of an atom led to its splitting and emitting more particles that bombarded the other atoms. This was carried out in 1942, and led to the development of the A-bomb shortly after.

ALPHA PARTICLES

Alpha radiation (or alpha rays) was distinguished and named by E. R. Rutherford in 1909, who found by measuring the charge and mass of alpha particles that they are the nuclei of ordinary helium atoms. Alpha particles consist of two protons and two neutrons.

BETA PARTICLES

Beta radiation (or beta rays) was identified and named by E. Rutherford too. He found that beta rays consists of high-speed electrons. Unlike alpha and gamma particles, whose energy can be explained as the difference of the energies of the radioactive nucleus before and after emission, beta particles emerge with a variable energy. This apparent violation of the law of conservation of energy led to the hypothesis that a second undetected particle, the neutrino, is emitted along with the electron and shares the total available energy. In some forms of induced, or artificial, radioactivity, the electron's antiparticle, the positron, is emitted from the excited nucleus; the positron in this case is also called a beta particle and denoted by b^+ (the ordinary beta particle is b^-).

GAMMA RADIATION

Gamma radiation is high-energy photons emitted as one of the three types of radiation resulting from natural radioactivity. It is the most energetic form of electromagnetic radiation, with a very short wavelength (high frequency). Wavelengths of the longest gamma radiation are less than 10^{-10} m, with frequencies greater than 10^{18} hertz (cycles per sec). Gamma rays are essentially very energetic X rays; the distinction between the two is not based on their intrinsic nature but rather on their origins.

X rays are emitted during atomic processes involving energetic electrons. Gamma radiation is emitted by excited nuclei or other processes involving subatomic particles; it often accompanies alpha or beta radiation, as a nucleus emitting those particles may be left in an excited (higher-energy) state. The applications of gamma radiation are much the same as those of X rays, both in medicine and in industry. In medicine, gamma ray sources are used for cancer treatment and for diagnostic purposes.

Some gamma-emitting radioisotopes are also used as tracers. In industry, principal applications include inspection of castings and welds. Data from artificial satellites and high-altitude balloons have indicated that a flux of gamma radiation is reaching the earth from outer space, thus opening up the field of research known as gamma-ray astronomy.

DECAY

ALPHA DECAY

There are three types of radioactive decay, named by the type of particle they produced. All radioactive decay is the result of an unstable nucleus and release energy. The first is the alpha decay, which emits an alpha particle, a positively charged and heavy particle that does not penetrate too well. It was shown that the alpha particle is actually a Helium nucleus, consisting of two protons and two neutrons. When the unstable nucleus emits the alpha particle, the atomic number of the atom decreases by two and the atomic mass number decreases by four.

BETA DECAY

The beta particle is an electron. When beta decay occurs, a neutron in the nucleus converts into a proton and an electron. The electron is emitted and takes the released energy with it. When it does not carry enough enery, however, the remaining is carried by a small, mass less (or very light) neutral particle called the neutrino. The beta particle is therefore negative and more penetrating than the alpha particle.

GAMMA DECAY

The Gamma radiation is unaffected by magnetic field as it travels; therefore it is electrically neutral. It is actually a type of electromagnetic radiation, and not a particle of significant mass. Gamma rays is one of the most energetic types of electromagnetic radiation, because its wavelength is very short and its frequency very high, (they are therefore invisible to the eye).

Gamma radiation often accompanies alpha decay or beta decay, since the nucleus emitting those particles may still be excited, and the gamma ray is the result of the extra energy. Gamma rays are the most penetrating type of radiation.

NATURAL GROWTH

Radioactive decay obeys the rules of natural (exponential) growth or decay. That is, the rate of decay is directly proportional to the amount of undecayed substance present. (Chemically, the rate of reaction is directly proportional to the amount of unreacted reactant.) Thus, the first derivative of the substance, say, Uranuim (U) with respect to time is proportional to the amount of undecayed U present.

$$d[U]/dt = k[U].$$

Solving this gives

$$d[U]/[U] = kdt, \text{ or}$$

$$\ln[U] + C = kt,\ [U] = a^*e(kt),$$

where a is calculated to be the amounf or Uranium at the start of the decay.

A proporty of the exponential graph is that starting at any arbitrarily chosen time, the time it takes for the amount of undecayed Uranium to decrease by half (or any fraction) is the same. This period of time, the time it take for half the sample to undergo radioactive decay, is called the Half-life of that element.

After one half-life, the amount of unreacted sample left is alway one half the amount of unreacted sample before that period of time. Thus, if the half life of a particular sample is 10 years, then after 20 years one-quarter of undecayed sample will be left.

INDUCED RADIOACTIVITY

The induced radioactivity was discovered by French physicists Iren Joliot-Curie (1897-1956) and Frederic Joliot-Curie (1900-1960). The first synthetic nuclear reaction was realized by Rutherford. It is possible to register the charged particles which are taking off at a radioactive decay. It allows to use atoms of the synthetic radioactive isotopes.

NUCLEAR REACTIONS

Nuclear reactions. There are 2 stages of nuclear reactions. The first stage of nuclear reaction consists in entrapment by a kern of a particle. For a small time interval the energy is uniformly proportioned between all particles of a compound nucleus. The second stage of nuclear response (start of a particle from a compound nucleus) occurs as a result of casual diversions from a uniform distribution of energy.

There is a time between entrapment of particles and their start from a kern, which is significant more then nuclear time $t_{n} = 10^{-22}$ s. All process of nuclear reaction is possible to figure by the formula:

$$_{Z1}X^{A1} + a \rightarrow _{Z2}Y^{A2} \rightarrow _{Z3}C^{A3} + b$$

$_{Z1}X^{A1}$ -initial kern; a, b -particles;

$_{Z2}X^{A2}$ -compound nucleus; $_{Z3}X^{A3}$ -terminating kern

DIVISION OF KERNS

The heavy kerns are labile. The kern, which was exited as a result of a neutron capture, can be divided into identical parts. The nuclear fission on two debrises is accompanied by eduction of huge energy. The debrises, which are formed at division, are radioactive and can emit neutrons. As the result the exuberant number of neutrons is formed, that gives in radio-activity of yields of decay.

REACTION OF DIVISION

CHAIN REACTION OF DIVISION

The division of kerns of uranium occurs at interaction to prompt or sluggish neutrons, but the irradiation by sluggish neutrons is more effective. The practical interest is represented by reactions, at which the major energy is oozed and there are neutrons, which give in magnification of number of the acts of division. For realization of chain reaction the maintenance of fixed number of neutrons is necessary.

For an estimation of number of neutrons is inlet coefficient of manifolding of neutrons k. It determines the ratio of number of neutrons, which arise at reaction, to number of neutrons, which participated in reaction. If $k \geq 1$, the reaction will be explicated. The minimum sizes of a fissile region, at which the realization of a chain reaction is possible, term as the critical sizes. The minimum quantity of a fissionable material necessary to sustain a nuclear chain reaction is called the critical mass.

CONTROLLABLE REACTION OF DIVISION

The controllable reactions occurs in a spacial nuclear reactors. Reactors can be used for research or for power production. A research reactor is designed to produce various beams of radiation for experimental application; the heat produced is a waste product and is dissipated as efficiently as possible.

In a power reactor the heat produced is of primary importance for use in driving conventional heat engines; the beams of radiation are controlled by shielding.

A-BOMB

Practical fissionable nuclei for atomic bombs are the isotopes uranium-235 and plutonium-239, which are capable of undergoing chain reaction. If the mass of the fissionable material exceeds the critical mass (a few pounds), the chain reaction multiplies rapidly into an uncontrollable release of energy.

An atomic bomb is detonated by bringing together very rapidly (e.g., by means of a chemical explosive) two sub critical masses of fissionable material, the combined mass exceeding the critical mass.

An atomic bomb explosion produces, in addition to the shock wave accompanying any explosion, intense neutron and gamma radiation, both of which are very damaging to living tissue.

The neighborhood of the explosion becomes contaminated with radioactive fission products. Some radioactive products are borne into the upper atmosphere as dust or gas and may subsequently be deposited partially decayed as radioactive fallout far from the site of the explosion.

ELEMENTARY PARTICLES

All subatomic particles, such as nucleons or electrons, and the particles they release when undergoing processes such as radioactive decay, are organized into classes.

All particles that obey the Pauli Exclusion Principle (no two particles can have the same four quantum numbers in a system, such as an atom) are called fermions because the Fermi-Dirac statistics apply to them. All particles that don't obey the principle are called bosons because the Bose-Einstein statistics apply to them.

FERMIONS

Fermions have half-integral spin (intrinsic angular momentum). They are deemed anitsocial particles because they obey the exculsion principle, meaning that no two fermion with the same spin can be found in a system. The antiparticles for fermions have the same mass but opposite charge and magnetic moment as the corresponding particle.

BARYONS

The first type of fermion is the baryon. They are those particles that experience the strong nuclear force. They consist of three quarks (quark triplet). The most notable baryons are the nucleons, the two states of which are protons and neutrons, the particles that make up the nucleus.

Other, heavier baryons, called hyperons also exist, but they are unstable, and usually decays to a form of nucleon(heavier baryons include the sigma, the lamda and the delta particles). Since all baryons decay to the nucleon form, the nucleons are known as the baryon ground state.

All baryons obey the law of baryon family numbers (the family number is +1 for particles such as protons and -1 for antiparticles such as antiprotons), which states that the sum of the family number of all reactants in an interaction is equal to the sum of the family number of all products.

This means that, for every particle created or destroyed, a corresponding antiparticle is also created or destroyed. For example, a proton colliding with an antiproton will result in the mutual destruction of both particles, or, a proton and an antiproton can be created simultaneously.

LEPTONS

Leptons are fermions that do not experience the strong nuclear force. They are only weakly interactive, and can be created by nuclear decay. Leptons include such particles as electrons, muons, neutrinos and the tau particle. So, for example, an electron and a neutrino can be created by the decay of an unstable neutron to a proton. As fermions they obey the exclusion principle, so no two electrons can occupy the same energy level and orbital of an atom (a system) and have the same spin. Leptons are the lightest fermions.

There are three separate laws concerning the conservation of lepton family numbers (the family numbers for leptons, like for baryons, is +1 for particles and -1 for antiparticles), for the electrons, muons and taus.

BOSONS

Bosons are particles that, unlike the fermions, do not obey the exclusion principle. Generally, fermions are particles that compose atomic structure but bosons carry out the interaction (forces) between fermions. Bosons have integral spin and are deemed social because many bosons can by placed in the same system in the same energy state at once.

MESONS

Mesons are a class of bosons whose mass falls between the lighter lepton fermions and the heavier baryon fermions. They consist of two quarks (quark pair), and include such particles as the p meson (pion) and the *K* meson (kaon). It was previously thought that the muon was a type of meson (m meson) because of its large mass (about 200 times that of the electron), but the behaviour of the muon doesn't fit into the description of mesons, so it was reclassified.

The pion is accountable for carrying out the strong nuclear force, which is the force that holdes the nucleus together, despite tmememndous electric repulsion.

OTHER BOSONS

Other bosons include the W and Z particles are responsible for carrying out the weak nuclear force and wjocj are about 200 heavier than the proton, and the photon, the particle form of light. The photon is its own antiparticple, as is the neutral pion.

Chapter 5

Ionizing Radiation

IONIZING VERSUS NON-IONIZING RADIATION

We live in a sea of radiation. In recent years, people have learned to fear the effects of radiation. They don't want to live near nuclear reactors. They are frightened by reports of links between excess exposure to sunlight and skin cancer. They are afraid of the leakage from microwave ovens, or the radiation produced by their television sets.

Several factors combine to heighten the public's anxiety about both the short-range and long-range effects of radiation. Perhaps the most important source of fear is the fact that radiation can't be detected by the average person. Furthermore, the effects of exposure to radiation might not appear for months or years or even decades.

To understand the biological effects of radiation we must first understand the difference between ionizing radiation and non-ionizing radiation. In general, two things can happen when radiation is absorbed by matter: excitation or ionization.

- Excitation occurs when the radiation excites the motion of the atoms or molecules, or excites an electron from an occupied orbital into an empty, higher-energy orbital.
- Ionization occurs when the radiation carries enough energy to remove an electron from an atom or molecule.

Because living tissue is 70-90 per cent water by weight, the dividing line between radiation that excites electrons and radiation that forms ions is often assumed to be equal to the ionization of water: 1216 kJ/mol. Radiation that carries less energy can only excite the water molecule. It is therefore called non-ionizing radiation. Radiation that carries more energy than 1216 kJ/mol can remove an electron from a water molecule, and is therefore called ionizing radiation.

The table below contains estimates of the energies of various kinds of radiation. Radio waves, microwaves, infrared radiation, and visible light are all forms of non-ionizing radiation.

X-rays, γ-rays, and α- and β -particles are forms of ionizing radiation. The dividing line between ionizing and non-ionizing radiation in the electromagnetic spectrum falls in the ultraviolet portion of the spectrum. It is therefore useful to divide the UV spectrum into two categories: UV_A and UV_B. Radiation at the high-energy end of the UV spectrum can be as dangerous as x-rays or γ-rays.

Energies of Ionizing and Non-Ionizing Forms of Radiation

Radiation		*Typical Frequency (s^{-1})*		*Typical Energy (kJ/mol)*		
Particles						
α-particles				4.1×10^8		
β -particles				1.5×10^7		
Electromagnetic Radiation						Ionizing
Cosmic rays		6×10^{21} s^{-1}		2.4×10^9		Radiation
γ-rays		3×10^{20} s^{-1}		1.2×10^8		
x-rays		3×10^{17} s^{-1}		1.2×10^5		
Ultraviolet		3×10^{15} s^{-1}		1200		
Visible		5×10^{14} s^{-1}		200		
Infrared		3×10^{13} s^{-1}		12		Non-ionizing
Microwaves		3×10^9 s^{-1}		1.2×10^{-3}		Radiation
Radio waves		3×10^7 s^{-1}		1.2×10^{-5}		

When ionizing radiation passes through living tissue, electrons are removed from neutral water molecules to produce H_2O^+ ions. Between three and four water molecules are ionized for every 1.6×10^{-17} joules of energy absorbed in the form of ionizing radiation.

$$H_2^O \rightarrow H_2^{O^+} + e^-$$

The H_2O^+ ion should not be confused with the H_3O^+ ion produced when acids dissolve in water. The H_2O^+ ion is an example of a free radical, which contains an unpaired valence-shell electron. Free radicals are extremely reactive. The radicals formed when ionizing radiation passes through water are among the strongest oxidizing agents that can exist in aqueous solution.

At the molecular level, these oxidizing agents destroy biologically active molecules by either removing electrons or removing hydrogen atoms. This often leads to damage to the membrane, nucleus, chromosomes, or mitochondria of the cell that either inhibits cell division, results in cell death, or produces a malignant cell.

BIOLOGICAL EFFECTS OF IONIZING RADIATION

From the time that radioactivity was discovered, it was obvious that it caused damage. As early as 1901, Pierre Curie discovered that a sample of radium placed on his skin produced wounds that were very slow to heal. What some find surprising is the magnitude of the difference between the biological effects of non-ionizing radiation, such as light and microwaves, and ionizing radiation, such as high-energy ultraviolet radiation, x-rays, γ-rays, and α- or β -particles.

Radiation at the low-energy end of the electromagnetic spectrum, such as radio waves and microwaves, excites the movement of atoms and molecules, which is equivalent to heating the sample.

Radiation in or near the visible portion of the spectrum excites electrons into higher-energy orbitals. When the electron eventually falls back to a lower-energy state, the excess energy is given off to neighboring molecules in the form of heat. The principal effect of non-ionizing radiation is therefore an increase in the temperature of the system.

We experience the fact that biological systems are sensitive to heat each time we cook with a microwave oven, or spend too long in the sun. But it takes a great deal of non-ionizing radiation to reach dangerous levels. We can assume, for example, that absorption of enough radiation to produce an increase of about 6C in body temperature would be fatal. Since the average 70-kilogram human is 80 per cent water by weight, we can use the heat capacity of water to calculate that it would take about 1.5 million joules of non-ionizing radiation to kill the average human. If this energy was carried by visible light with a frequency of 5×10^{14} s^{-1}, it would correspond to absorption of about seven moles of photons.

Ionizing radiation is much more dangerous. A dose of only 300 joules of x-ray or α-ray radiation is fatal for the average human, even though this radiation raises the temperature of the body by only 0.001C. α-particle radiation is even more dangerous; a dose equivalent to only 15 joules is fatal for the average human. Whereas it takes seven moles of photons of visible light to produce a fatal dose of non-ionizing radiation, absorption of only 7×10^{-10} moles of the α-particles emitted by ^{238}U is fatal.

There are three ways of measuring ionizing radiation.

- Measure the activity of the source in units of disintegrations per second or curies, which is the easiest measurement to make.
- Measure the radiation to which an object is exposed in units of roentgens by measuring the amount of ionization produced when this radiation passes through a sample of air.
- Measure the radiation absorbed by the object in units of radiation absorbed doses or "rads". This is the most useful quantity, but it is the hardest to obtain.

One radiation absorbed dose, or rad, corresponds to the absorption of 10^{-5} joules of energy per gram of body weight. Because this is equivalent to 0.01 J/kg, one rad produces an increase in body temperature of about 2×10^{-6}C. At first glance, the rad may seem to be a negligibly small unit of measurement. The destructive power of the radicals produced when water is ionized is so large, however, that cells are inactivated at a dose of 100 rads, and a dose of 400 to 450 rads is fatal for the average human.

Not all forms of radiation have the same efficiency for damaging biological organisms. The faster energy is lost as the radiation passes through the tissue, the more damage it does. To correct for the differences in radiation biological effectiveness (RBE) among various forms of radiation, a second unit of absorbed dose has been defined. The roentgen equivalent man, or rem, is the absorbed dose in rads times the biological effectiveness of the radiation.

$$\text{rems} = \text{rads} \times \text{RBE}$$

Values for the RBE of different forms of radiation are given in the table below.

Table. The Radiation Biological Effectiveness of Various Forms of Radiation

Radiation		*RBE*
x-rays and γ-rays		1
β - particles with energies larger than 0.03 MeV		1
β - particles with energies less than 0.3 MeV		1.7
Thermal (slow-moving) neutrons		3
Fast-moving neutrons or protons		10
α - particles or heavy ions		20

Estimates of the per capita exposure to radiation in the United States are summarized in the table below. These estimates include both external and internal sources of natural background radiation.

Table. Average Whole-Body Exposure Levels for Sources of Ionizing Radiation

Source		*Per Capita Dose (rems / y)*
Natural background		0.082
Medical x-rays		0.077
Nuclear test fallout		0.005
Consumer and industrial products		0.005
Nuclear power industry		0.001
	total:	0.170

External sources include cosmic rays from the sun and α-particles or γ-rays emitted from rocks and soil. Internal sources include nuclides that enter the body when we breathe (^{14}C, ^{85}Kr, ^{220}Rn, and ^{222}Rn) and through the food chain (^{3}H, ^{14}C, ^{40}K, ^{90}Sr, ^{131}I, and ^{137}Cs). The actual dose from natural radiation depends on where one lives. People who live in the Rocky Mountains, for example, receive twice as much background radiation as the national average because there is less atmosphere to filter out the cosmic rays from the sun.

The average dose from medical x-rays has decreased in recent years because of advances in the sensitivity of the photographic film used for x-rays. Radiation from nuclear test fallout has also decreased as a result of the atmospheric nuclear test ban.

The threat of fallout from the testing of nuclear weapons can be appreciated by noting that a Chinese atmospheric test in 1976 led to the contamination of milk in the Harrisburg, PA, vicinity at a level of 300 pCi (3.00×10^{-10} Ci) per lite. This was about eight times the level of contamination (41 pCi per liter) that resulted from the accident at Three Mile Island.

The contribution to the radiation absorbed dose from consumer and industrial products includes radiation from construction materials, x-rays emitted by television sets, and inhaled tobacco smoke. The most recent estimate of the total radiation emitted from the mining and milling of uranium, the fabrication of reactor fuels, the storage of radioactive wastes, and the operation of nuclear reactors is less than 0.001 rem per year.

The total dose from ionizing radiation for the average American is about 0.170 rem per year. The Committee on the Biological Effects of Ionizing Radiation of the National Academy of Sciences recently estimated that an increase in this dose to a level of 1 rem per year would result in 169 additional deaths from cancer per million people exposed.

This can be compared with the 170,000 cancer deaths that would normally occur in a population this size that was not exposed to this level of radiation. The principal effect of low doses of ionizing radiation is to induce cancers, which may take up to 20 years to develop. What is the effect of high doses of ionizing radiation? Cells that are actively dividing are more sensitive to radiation than cells that aren't. Thus, cells in the liver, kidney, muscle, brain, and bone are more resistant to radiation than the cells of bone marrow, the reproductive organs, the epithelium of the intestine, and the skin, which suffer the most damage from radiation.

Damage to the bone marrow is the main cause of death at moderately high levels of exposure (200 to 1000 rads). Damage to the gastrointestinal tract is the major cause of death for exposures on the order of 100 to 10,000 rads. Massive damage to the central nervous system is the cause of death from extremely high exposures (over 10,000 rads).

NATURAL VERSUS INDUCED RADIOACTIVITY

NATURAL RADIOACTIVITY

The vast majority of the nuclides found in nature are stable. If our planet is 4.6 billion years old, the only radioactive isotopes that should remain are members of three classes.

- Isotopes with half-lives of at least 10^9 years, such as ^{238}U.
- Daughter nuclides produced when long-lived radioactive nuclides decay, such as the ^{234}Th ($t_{1/2}$ = 24.1 days) produced by the α-decay of ^{238}U.
- Nuclides such as ^{14}C that are still being synthesized.

One factor that influences the stability of a nuclide is the ratio of neutrons to protons. (Nuclei that contain either too many or too few neutrons are unstable.) Another factor that affects the stability of nuclides can be understood by examining patterns in the numbers of protons and neutrons in stable nuclides.

Half of the elements in the periodic table must have an odd number of protons because atomic numbers that are odd are just as likely to occur as those that are even. In spite of this, about 80 per cent of the stable nuclides have an even number of protons. Very few elements with an odd atomic number have more than one stable isotope. Stable isotopes abound, however, among elements with even atomic numbers. Ten stable isotopes are known for tin (Z = 50), for example. It is also interesting to note that 91 per cent of the stable isotopes of elements with an odd number of protons have an even number of neutrons.

These observations suggest that certain combinations of protons and neutrons are particularly stable. There are magic numbers of electrons. Electron configurations with 2, 10, 18, 36, 54, and 86 electrons are unusually stable. There also seem to be magic numbers of neutrons and protons. Nuclei with 2, 8, 20, 28, 50, 82, or 126 protons or neutrons are unusually stable. This observation explains the anomalously large binding energies observed for ^{4}He, ^{16}O, and ^{20}Ne.

In each case, the nuclide has an even number of both protons and neutrons. ^{20}Ne has a magic number of nucleons when both protons and neutrons are counted. ^{4}He and ^{16}O have magic numbers of both protons and neutrons. The resulting stability of the ^{4}He nucleus might explain why so many heavy nuclei undergo α-particle decay by ejecting an $^{4}He^{2+}$ ion from the nucleus of the atom.

If nuclei tend to be more stable when they have even numbers of protons and neutrons, it isn't surprising that nuclides with an odd number of both protons and neutrons are unstable. ^{40}K is one of only five naturally occurring nuclides that contain both an odd number of protons and an odd number of neutrons.

This nuclide simultaneously undergoes the electron capture and positron emission expected for neutron-poor nuclides and the electron emission observed with neutron-rich nuclides. Only 18 radioactive isotopes with atomic numbers of 80 or less can be found in nature. With the exception of ^{14}C, which is continuously synthesized in the atmosphere, all these elements have lifetimes longer than 10^9 years.

Although these isotopes all undergo radioactive decay, they decay so slowly that reasonable quantities are still present, 4.6 billion years after the planet was formed. Another 45 natural radioactive isotopes have atomic numbers larger than 80. These nuclides fall into three families. The parent nuclide is ^{232}Th, which undergoes α-decay to form ^{228}Ra. The product of this reaction decays by β-emission to form ^{228}Ac, which decays to ^{228}Th, and so on, until the stable ^{208}Pb isotope is formed. This family of radionuclides is called the 4n series, because all its members have a mass number that can be divided by 4 A second family of radioactive nuclei starts with ^{238}U and decays to form the stable ^{206}Pb isotope. Every member of this series has a mass number that fits the equation 4n + 2.

The third family, known as the 4n + 3 series, starts with ^{235}U and decays to ^{207}Pb. A 4n + 1 series once existed, which started with ^{237}Np and decayed to form the only stable isotope of bismuth, ^{209}Bi.

The half-life of every member of this series is less than 2×10^6 years, however, so none of the nuclides produced by the decay of neptunium remain in detectable quantities on the earth.

INDUCED RADIOACTIVITY

In 1934, Irene Curie, the daughter of Pierre and Marie Curie, and her husband, Frederic Joliot, announced the first synthesis of an artificial radioactive isotope. They bombarded a thin piece of aluminum foil with -particles produced by the decay of polonium and found that the aluminum target became radioactive. Chemical analysis showed that the product of this reaction was an isotope of phosphorus.

$$^{27}_{13}Al + {}^{4}_{2}He \rightarrow {}^{30}_{15}P + {}^{1}_{0}n$$

In the next 50 years, more than 2000 other artificial radionuclides were synthesized. A shorthand notation has been developed for nuclear reactions such as the reaction discovered by Curie and Joliot. The parent (or target) nuclide and the daughter nuclide are separated by parentheses that contain the symbols for the particle that hits the target and the particle or particles released in this reaction.

$$^{27}_{13}Al(\alpha, n)\,{}^{30}_{15}P$$

The nuclear reactions used to synthesize artificial radionuclides are characterized by enormous activation energies. Three devices are used to overcome these activation energies: linear accelerators, cyclotrons, and nuclear reactors. Linear accelerators or cyclotrons can be used to excite charged particles such as protons, electrons, α-particles, or even heavier ions, which are then focused on a stationary target. The following reaction, for example, can be induced by a cyclotron or linear accelerator.

$$^{24}_{12}Mg + {}^{2}_{1}H \rightarrow {}^{22}_{11}Na + {}^{4}_{2}He$$

Because these reactions involve the capture of a positively charged particles, they usually produce a neutron-poor nuclide. Artificial radionuclides are also synthesized in nuclear reactors, which are excellent sources of slow-moving, or thermal neutrons.

The absorption of a neutron usually results in a neutron-rich nuclide. The following neutron absorption reaction occurs in the cooling systems of nuclear reactors cooled with liquid sodium metal.

$$^{23}_{11}Na + {}^{1}_{0}n \rightarrow {}^{24}_{11}Na + \gamma$$

In 1940, absorption of thermal neutrons was used to synthesize the first elements with atomic numbers larger than the heaviest naturally occurring element, uranium. The first of these truly artificial elements were neptenium and plutonium, which were synthesized by McMillan and Abelson by irradiating ^{238}U with neutrons to form ^{239}U

$$^{238}_{92}U + {}^{1}_{0}n \rightarrow {}^{239}_{92}U + \gamma \quad (neutron\,capture)$$

which undergoes β-decay to form ^{239}Np and then ^{239}Pu.

$$^{239}_{92}U \rightarrow {}^{239}_{93}Np + {}^{0}_{-1}e$$

$$^{239}_{93}Np \rightarrow {}^{239}_{94}Pu + {}^{0}_{-1}e$$

Larger bombarding particles were eventually used to produce even heavier transuranium elements.

$$^{253}_{99}Es + ^{4}_{2}He \rightarrow ^{256}_{101}Md + ^{1}_{0}n$$

$$^{246}_{96}Cm + ^{12}_{6}C \rightarrow ^{254}_{102}No + 4\,^{1}_{0}n$$

The half-lives for α-decay and spontaneous fission decrease as the atomic number of the element increases. Element 104, for example, has a half-life for spontaneous fission of 0.3 seconds.

Elements therefore become harder to characterize as the atomic number increases.

Recent theoretical work has predicted that a magic number of protons might exist at Z = 114.

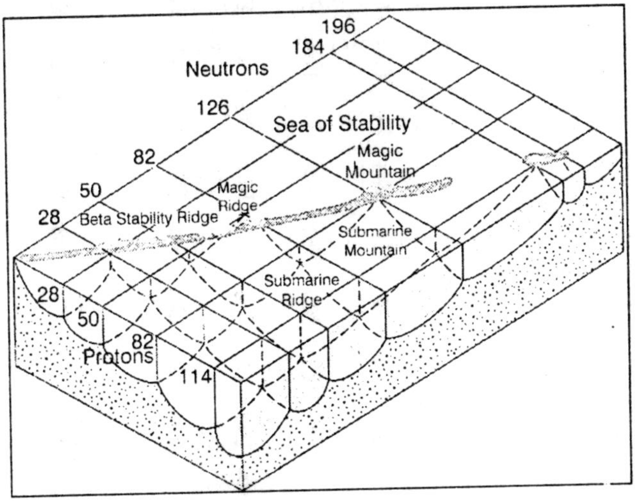

Fig. Unstable Nuclides

This work suggests that there is an island of stability in the sea of unstable nuclides.

If this theory is correct, superheavy elements could be formed if we could find a way to cross the gap between elements Z = 109 through Z = 114.

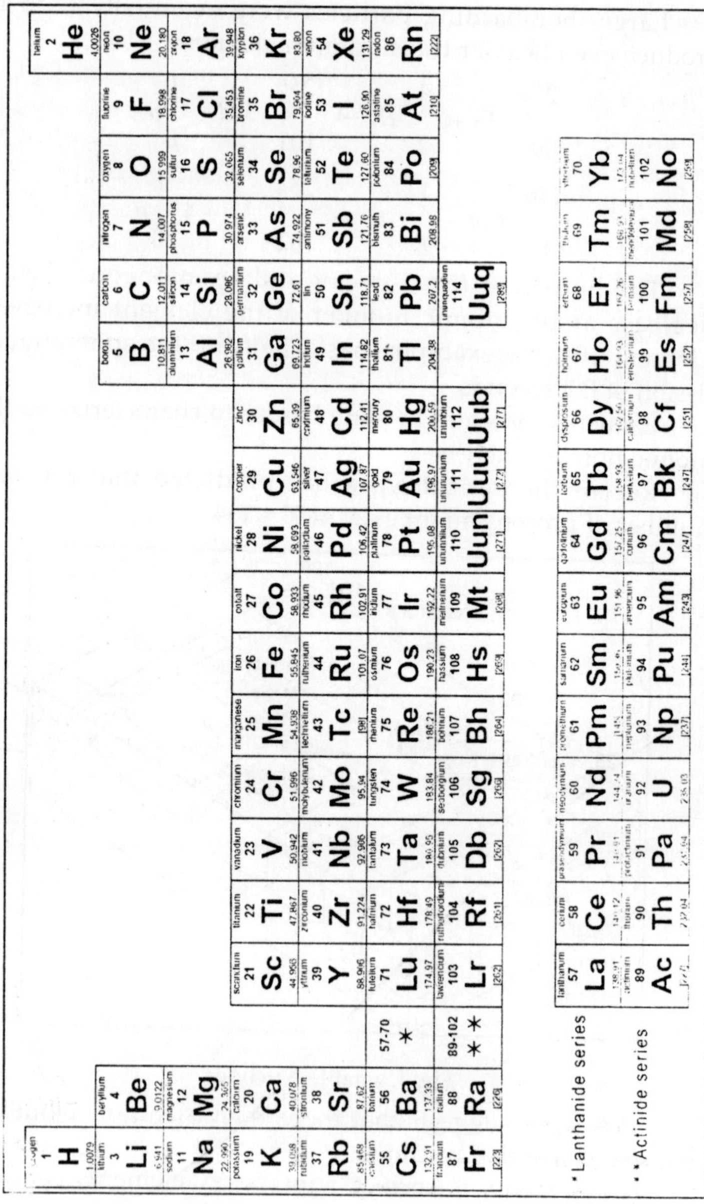

Fig. An Extended Version of the Periodic Table that Predicts the Positions for all Elements.

There is some debate about the number of neutrons needed to overcome the proton-proton repulsion in a nucleus with 114 protons. The best estimates suggest that at least 184 neutrons, and perhaps as many as 196, would be needed. It is not an easy task to bring together two particles that give both the correct number of total protons and the necessary neutrons to produce a nuclide with a half-life long enough to be detected.

If we start with a relatively long-lived parent nuclide, such as ^{251}Cf ($t_{1/2}$ = 800 y) and bombard this nucleus with a heavy ion, such as ^{32}S, we can envision producing a daughter nuclide with the correct atomic number, but the mass number would be too small by at least 16 amu.

$$^{251}_{98}Cf + ^{32}_{16}S \rightarrow ^{282}_{114}X + ^{1}_{0}n$$

An expanded periodic table for elements up to Z = 168 is shown in the figure below. Elements 104 through 112 are transition metals that fill the 6d orbitals. Elements 113 through 120 are main-group elements in which the 7p and 8s orbitals are filled.

The next subshell is the 5g atomic orbital, which can hold up to 18 electrons. There is reason to believe that the 5g and 6f orbitals will be filled at the same time. The next 32 elements are therefore grouped into a so-called superactinide series.

NUCLEAR FISSION AND NUCLEAR FUSION

NUCLEAR FISSION

The graph of binding energy per nucleon suggests that nuclides with a mass larger than about 130 amu should spontaneously split apart to form lighter, more stable, nuclides. Experimentally, we find that spontaneous fission reactions occur for only the very heaviest nuclides those with mass numbers of 230 or more.

Even when they do occur, these reactions are often very slow. The half-life for the spontaneous fission of ^{238}U, for example, is 10^{16} years, or about two million times longer than the age of our planet!

We don't have to wait, however, for slow spontaneous fission reactions to occur. By irradiating samples of heavy nuclides with slow-moving thermal neutrons it is possible to induce fission reactions. When ^{235}U absorbs a thermal neutron, for example, it splits into two particles of uneven mass and releases an average of 2.5 neutrons, as shown in the figure below.

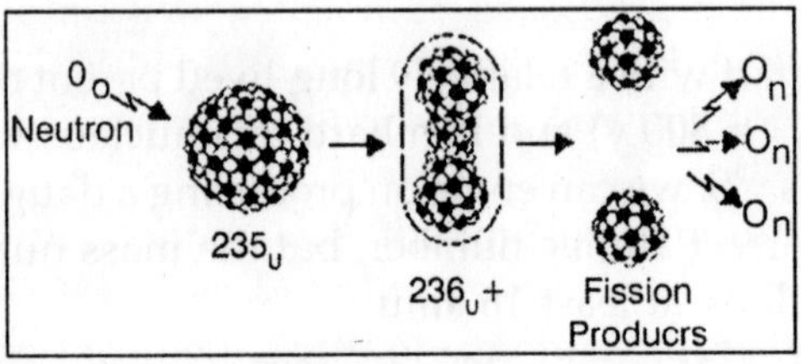

Fig. The Absorption of a Neutron by ^{238}U Induces Oscillationsin the Nucleus that Deform it Until it Splits into Fragments theWay a Drop of Liquid Might Break into Smaller Droplets.

More than 370 daughter nuclides with atomic masses between 72 and 161 amu are formed in the thermal-neutron-induced fission of ^{235}U, including the two products shown below.

$$^{235}_{92}U + ^{1}_{0}n \rightarrow ^{139}_{56}Ba + ^{94}_{36}Kr + 3\,^{1}_{0}n$$

Several isotopes of uranium undergo induced fission. But the only naturally occurring isotope in which we can induce fission with thermal neutrons is ^{235}U, which is present at an abundance of only 0.72 per cent. The induced fission of this isotope releases an average of 200 MeV per atom, or 80 million kilojoules per gram of ^{235}U. The attraction of nuclear fission as a source of power can be understood by comparing this value with the 50 kJ/g released when natural gas is burned.

The first artificial nuclear reactor was built by Enrico Fermi and co-workers beneath the University of Chicago's football stadium and brought on line on December 2, 1942. This reactor, which produced several kilowatts of power, consisted of a pile of graphite blocks weighing 385 tons stacked in layers around a cubical array of 40 tons of uranium metal and uranium oxide. Spontaneous fission of ^{238}U or ^{235}U in this reactor produced a very small number of neutrons.

But enough uranium was present so that one of these neutrons induced the fission of a ^{235}U nucleus, thereby releasing an average of 2.5 neutrons, which catalyzed the fission of additional ^{235}U nuclei in a chain reaction, as shown in the figure below. The amount of fissionable material necessary for the chain reaction to sustain itself is called the critical mass.

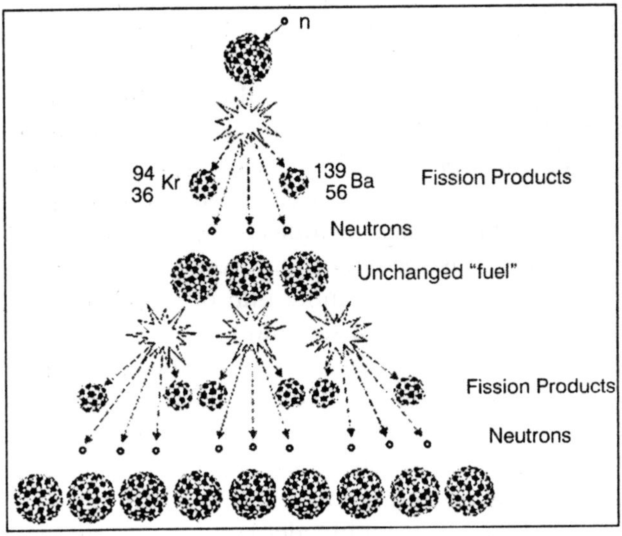

Fig. Fission Reaction.

The Fermi reactor at Chicago served as a prototype for larger reactors constructed in 1943 at Oak Ridge, Tennessee, and Hanford, Washington, to produce ^{239}Pu for one of the atomic bombs dropped on Japan at the end of World War II. As we have seen, some of the neutrons released in the chain reaction are absorbed by ^{238}U to form ^{239}U, which undergoes decay by the successive loss of two β -particles to form ^{239}Pu. ^{238}U is an example of a fertile nuclide.

It doesn't undergo fission with thermal neutrons, but it can be converted to ^{239}Pu, which does undergo thermal-neutron-induced fission. Fission reactors can be designed to handle naturally abundant ^{235}U, as well as fuels described as slightly enriched (2-5 per cent ^{235}U), highly enriched (20-30 per cent ^{235}U), or fully enriched (> 90 per cent ^{235}U). Heat generated in the reactor core is transferred to a cooling agent in a closed system. The cooling agent is then passed through a series of heat exchangers in which water is heated to steam.

The steam produced in these exchangers then drives a turbine that generates electrical power. There are two ways of specifying the power of such a plant: the thermal energy produced by the reactor or the electrical energy generated by the turbines. The electrical capacity of the plant is usually about one-third of the thermal power.

It takes 10^{11} fissions per second to produce one watt of electrical power. As a result, about one gram of fuel is consumed per day per megawatt of electrical energy produced. This means that one gram of waste products is produced per megawatt per day, which includes 0.5 grams of ^{239}Pu. These waste products must be either reprocessed to generate more fuel or stored for the tens of thousands of years it takes for the level of radiation to reach a safe limit.

If we could design a reactor in which the ratio of the ^{239}Pu or ^{233}U produced to the ^{235}U consumed was greater than 1, the reactor would generate more fuel than it consumed. Such reactors are known as breeders, and commercial breeder reactors are now operating in France.

The key to an efficient breeder reactor is a fuel that gives the largest possible number of neutrons released per neutron absorbed. The breeder reactors being built today use a mixture of PuO_2 and UO_2 as the fuel and fast neutrons to activate fission. Fast neutrons carry an energy of at least several KeV and therefore travel 10,000 or more times faster than thermal neutrons. ^{239}Pu in the fuel assembly absorbs one of these fast neutrons and undergoes fission with the release of three neutrons. ^{238}U in the fuel then captures one of these neutrons to produce additional ^{239}Pu.

The advantage of breeder reactors is obviousthey mean a limitless supply of fuel for nuclear reactors. There are significant disadvantages, however. Breeder reactors are more expensive to build. They are also useless without a subsidiary industry to collect the fuel, process it, and ship the ^{239}Pu to new reactors.

It is the reprocessing of ^{239}Pu that concerns most of the critics of breeder reactors. ^{239}Pu is so dangerous as a carcinogen that the nuclear industry places a limit on exposure to this material that assumes workers inhale no more than 0.2 micrograms of plutonium over their lifetimes. There is also concern that the ^{239}Pu produced by these reactors might be stolen and assembled into bombs by terrorist organizations.

The fate of breeder reactors in the United States is linked to economic considerations. Because of the costs of building these reactors and safely reprocessing the ^{239}Pu produced, the breeder reactor becomes economical only when the scarcity of uranium drives its price so high that the breeder reactor becomes cost effective by comparison.

If nuclear energy is to play a dominant role in the generation of electrical energy in the 21st century, breeder reactors eventually may be essential.

Although the "pile" Fermi constructed at the University of Chicago in 1942 was the first artificial nuclear reactor, it was not the first fission reactor to exist on Earth. In 1972, a group of French scientists discovered that uranium ore from a deposit in the Oklo mine in Gabon, West Africa, contained 0.4 per cent ^{235}U instead of the 0.72 per cent abundance found in all other sources of this ore.

Analysis of the trace elements in the ore suggested that the amount of ^{235}U in this ore was unusually small because natural fission reactors operated in this deposit for a period of 600,000 to 800,000 years about 2 billion years ago.

NUCLEAR FUSION

The graph of binding energy per nucleon suggests another way of obtaining useful energy from nuclear reactions. Fusing two light nuclei can liberate as much energy as the fission of ^{235}U or ^{239}Pu. The fusion of four protons to form a helium nucleus, two positrons (and two neutrinos), for example, generates 24.7 MeV of energy.

$$4\,{}^{1}_{1}H \rightarrow {}^{4}_{2}He + 2\,{}^{0}_{+1}e$$

Most of the energy radiated from the surface of the sun is produced by the fusion of protons to form helium atoms within its core.

Fusion reactions have been duplicated in man-made devices. The enormous destructive power of the ^{235}U-fueled atomic bomb dropped on Hiroshima on August 6, 1945, which killed 75,000 people, and the ^{239}Pu-fueled bomb dropped on Nagasaki three days later touched off a violent debate after World War II about the building of the next superweapon a fusion, or "hydrogen," bomb. Alumni of the Manhattan project, who had developed the atomic bomb, were divided on the issue.

Ernest Lawrence and Edward Teller fought for the construction of the fusion device. J. Robert Oppenheimer and Enrico Fermi argued against it. The decision was made to develop the weapon, and the first artificial fusion reaction occurred when the hydrogen bomb was tested in November 1952.

The history of fusion research is therefore the opposite of fission research. With fission, the reactor came first, and then the bomb was built. With fusion, the bomb was built long before any progress was made toward the construction of a controlled fusion reactor. More than 40 years after the first hydrogen bomb was exploded, the feasibility of controlled fusion reactions is still open to debate. The reaction that is most likely to fuel the first fusion reactor is the thermonuclear d-t, or deuterium-tritium, reaction. This reaction fuses two isotopes of hydrogen, deuterium (^{2}H) and tritium (^{3}H), to form helium and a neutron.

$${}^{2}_{1}H + {}^{3}_{1}H \rightarrow {}^{4}_{2}He + {}^{1}_{0}n$$

If we consider the implications of this reaction we can begin to understand why it is called a thermonuclear reaction and why it is so difficult to produce in a controlled manner. The d-t reaction requires that we fuse two positively charged particles. This means that we must provide enough energy to overcome the force of repulsion between these particles before fusion can occur. To produce a self-sustaining reaction, we have to provide the particles with enough thermal energy so that they can fuse when they collide.

Each fusion reaction is characterized by a specific ignition temperature, which must be surpassed before the reaction can occur. The d-t reaction has an ignition temperature above 10^8 K. In a hydrogen bomb, a fission reaction produced by a small atomic bomb is used to heat the contents to the temperature required to initiate fusion. Obtaining the same result in a controlled reaction is much more challenging.

Any substance at temperatures approaching 10^8 K will exist as a completely ionized gas, or plasma. The goals of fusion research at present include the following.

- To achieve the required temperature to ignite the fusion reaction.
- To keep the plasma together at this temperature long enough to get useful amounts of energy out of the thermonuclear fusion reactions.
- To obtain more energy from the thermonuclear reactions than is used to heat the plasma to the ignition temperature.

These are not trivial goals. The only reasonable container for a plasma at 10^8 K is a magnetic field. Both doughnut-shaped (toroidal) and linear magnetic bottles have been proposed as fusion reactors.

But reactors that produce high enough temperatures for ignition are not the same as the reactors that have produced long enough confinement times for the plasma to provide useful amounts of energy.

A second approach to a controlled fusion reactor involves hitting fuel pellets containing the proper reagents for the thermonuclear reaction with pulsed beams of laser power. If enough power was delivered, the fuel pellets would collapse upon themselves, or implode, to reach densities several orders of magnitude greater than normal. This could produce a plasma both hot enough and dense enough to initiate fusion reactions.

NUCLEAR SYNTHESIS AND NUCLEAR MEDICINE

NUCLEAR SYNTHESIS

Within the last 60 years, scientists have generated a new story of creation that offers a model for understanding how nuclei are synthesized. In 1929, Edwin Hubble provided evidence to suggest that our universe is expanding.

Between 1946 and 1948, George Gamow and co-workers generated a model which assumed that the primordial substance, or ylem, from which all other matter was created was an extraordinarily hot, dense singularity that exploded in a "Big Bang" and has been expanding ever since.

This model assumes that neutrons in the ylem were transformed into protons by β -decay. Neutrons and protons then combined to form ^{4}He atoms before the temperature and pressure of the fireball decayed to the point at which no further nuclear reactions were possible.

The first-generation stars condensed out of this cloud of hydrogen and helium. As the gas condensed by gravitational attraction, it became warmer. Eventually, the temperature reached 10^7 K, and first-generation, main-sequence stars were born. The temperature at the cores of these stars was high enough to ignite the following thermonuclear reactions.

$$^{1}_{1}H + {}^{1}_{1}H \rightarrow {}^{2}_{1}H + {}^{0}_{+1}e$$

$$^{1}_{1}H + {}^{2}_{1}H \rightarrow {}^{3}_{2}He$$

$$^{3}_{2}He + {}^{3}_{2}He \rightarrow {}^{4}_{2}He + 2\,{}^{1}_{1}H$$

The net result of these reactions is the formation of a helium atom from four protons.

$$4\,{}^{1}_{1}H \rightarrow {}^{4}_{2}He + 2\,{}^{0}_{+1}e$$

Eventually the heat generated in this reaction was enough to halt the gravitational collapse of the star, which entered a stable period during which the energy generated by this reaction balanced the energy radiated at the surface.

The hydrogen-burning reactions in a main-sequence star are concentrated in the core. When enough hydrogen has been consumed, the core begins to collapse, and the temperature of the core rises above 10^8 K. (The larger the star, the more rapidly it radiates energy from its surface, and the more rapidly it consumes the hydrogen in the core.)

As the core collapses, the hydrogen-containing outer shell expands, and the surface of the star cools. Stars that have reached this point in their evolution include the so-called red giants. Temperatures in the core of these red giants are high enough to ignite further thermonuclear fusion reactions, such as the following.

$${}^{4}_{2}He \rightarrow {}^{12}_{6}C$$

This reaction becomes the principal source of energy in a red giant, although there is undoubtedly some burning of hydrogen to helium in the outer shell of the star. As the amount of ^{12}C in the core increases, further reactions occur to form ^{16}O and ^{20}Ne.

$${}^{12}_{6}C + {}^{4}_{2}He \rightarrow {}^{16}_{8}O + \gamma$$

$${}^{16}_{8}O + {}^{4}_{2}He \rightarrow {}^{20}_{10}Ne + \gamma$$

Eventually, the helium in the core is exhausted, and the core collapses further, reaching temperatures of 6-7 × 10^8 K. At this point, more complex reactions take place that produce nuclides such as ^{28}Si and ^{32}S.

$$^{12}_{6}C + {}^{16}_{8}O \rightarrow {}^{28}_{14}Si + \gamma$$

$$^{16}_{8}O + {}^{16}_{8}O \rightarrow {}^{32}_{16}S + \gamma$$

Further gravitational collapse heats the core to temperatures above 10^9 K, and a complex sequence of reactions takes place to synthesize the nuclei with the highest binding energies, such as Fe and Ni.

If the star explodes in a supernova, its contents are ejected across space. Second-generation stars that condense in this region contain not only hydrogen and helium but elements with higher atomic number.

The best estimates of the age of the Milky Way suggest that our galaxy is about 15 billion years old. Our sun and its planets, however, are only 4.6 billion years old.

This suggests that the sun is a second-generation star. In such stars, the transformation of hydrogen to helium can be catalyzed by ^{12}C. Second-generation stars use a different mechanism to synthesis helium than first-generation stars. They use a sequence of reactions, such as the those shown here, that are catalyzed by isotopes heavier than helium.

Two processes can synthesize elements with atomic numbers larger than that of iron. One of them is relatively slow (s-process), the other is very rapid (r-process). Since the only way to synthesize nuclei with atomic numbers larger than iron is by the absorption of neutrons, both the s-process and the r-process result from (n,) reactions.

In the s-process, neutrons are captured one at a time to form a neutron-rich nuclide that has enough time to undergo α- or β-decay before another neutron can be absorbed. An example of an s-process sequence of reactions starts with ^{120}Sn. The capture of a neutron produces ^{121}Sn, which undergoes β-decay. If β-decay occurs before this nuclide captures another neutron, a stable isotope of antimony is formed.

Eventually, ^{121}Sb captures a neutron to produce ^{122}Sb, which is transformed into ^{122}Te by β -decay. ^{122}Te can undergo β -decay to form ^{122}I, or it can capture a neutron to form ^{123}Te. With ^{123}Te, we encounter a series of stable isotopes of tellurium. Thus, neutrons are slowly absorbed, one at a time, until we reach ^{127}Te, which decays to ^{127}I, the most abundant isotope of iodine.

$$ {}^{120}_{50}Sn(n,\gamma)\,{}^{121}_{50}Sn \xrightarrow{\beta^-} {}^{121}_{51}Sb(n,\gamma)\,{}^{122}_{51}Sb \xrightarrow{\beta^-} {}^{122}_{52}Te(5n,5\gamma)\,{}^{127}_{52}Te \xrightarrow{\beta^-} {}^{127}_{53}I $$

This slow process can't account for very heavy nuclides, such as ^{232}Th and ^{238}U, because the lifetimes of the intermediate nuclei with atomic numbers between 83 and 90 are too short for this step-by-step absorption of neutrons to proceed.

Synthesizing appreciable quantities of uranium and thorium requires a rapid process. In the r-process, a number of neutrons are captured in rapid succession, before there is time for α- or β -decay to take place.

Achieving an r-process reaction, however, requires a very high neutron flux. (These reactions occur during nuclear explosions, for example.) The neutron flux needed to fuel such reactions is not likely to occur in a normal star.

During the moment when a star explodes as a supernova, however, the conditions are ripe for r-process reactions. The heavier elements on this planet were therefore produced in a series of supernova explosions that occurred in this portion of the galaxy before our solar system condensed.

NUCLEAR MEDICINE

Ever since the first x-ray images were obtained by Roentgen in 1895, ionizing radiation and radio nuclides have played a vital role in medicine. This work has been so fruitful that a separate field known as nuclear medicine has developed.

Research in this field focuses on either therapeutic or diagnostic uses of radiation.

There are three standard approaches to fighting cancer: surgery, chemotherapy, and radiation. Surgery, by its very nature, is invasive. Chemotherapy and classic approaches to radiation therapy are not selective.

Research in recent years has therefore examined new approaches to radiation therapy that specifically attack tumor cells, without damaging normal tissue. The technique known as boron neutron capture therapy provides an example of this work.

Naturally occurring boron consists of two stable isotopes: ^{10}B (19.7 per cent) and ^{11}B (80.3 per cent). ^{10}B absorbs thermal neutrons to form ^{11}B in a nuclear excited state. Although ^{11}B in its nuclear ground state is stable, this excited ^{11}B nuclide undergoes fission to produce ^{7}Li and an α-particle.

$$^{10}_{5}B + {}^{1}_{0}n \rightarrow {}^{7}_{3}Li \; {}^{4}_{2}He$$

Because the energy of a thermal neutron is only about 0.025 eV, the neutrons that are not absorbed do relatively little damage to the normal tissue. The α-particle emitted in this reaction has an energy of 2.79 MeV, however, which makes it an extremely lethal form of radiation.

The RBE for α-particle radiation is larger than most other particles because this relatively massive particle loses energy very efficiently as it collides with matter. Radiation damage from the α-particle is therefore restricted to the immediate vicinity of the tissue that absorbed the thermal neutron.

Although other common nuclides in living tissue can absorb thermal neutrons, the ability of ^{10}B to absorb thermal neutrons is three orders of magnitude larger than these nuclides. The units with which this measurement is made can be understood by thinking about the area around the nuclide through which the neutron can pass and still be absorbed.

The neutron-capture cross section for a hydrogen atom corresponds to a circle with a radius of about 2×10^{-13} cm. For a nitrogen atom, the radius of this circle is about 10^{-12} cm. Boron has a neutron-capture cross section that would be described by a circle with a radius of about 2×10^{-9} cm. Virtually all of the neutron capture that occurs is therefore concentrated in the tissue that contains boron.

The potential of boron neutron capture therapy was recognized as early as 1936. Early clinical trials in the late 1950s and early 1960s failed to prolong the lives of patients suffering from brain tumors because the boron compounds available for testing did not concentrate selectively in the tumor cells.

Recent research has discovered boron-labeled compounds that are sufficiently selective for BNCT to become useful as an adjunct to surgery, or in place of surgery with inoperable cases.

Chapter 6

Nuclear Technology

Nuclear Technology is designed to extract usable energy from atomic nuclei via controlled nuclear reactions. The most common method today is through nuclear fission, though other methods include nuclear fusion and radioactive decay. All utility-scale reactors heat water to produce steam, which is then converted into mechanical work for the purpose of generating electricity or propulsion. In 2007, 14 per cent of the world's electricity came from nuclear power. More than 150 nuclear-powered naval vessels have been built, and a few radioisotope rockets have been produced.

USE OF NUCLEAR POWER

In 2007, nuclear's share of global electricity generation dropped to 14 per cent. According to the International Atomic Energy Agency, the main reason for this was an earthquake in western Japan on 16 July 2007, which shut down all seven reactors at the Kashiwazaki-Kariwa Nuclear Power Plant. There were also several other reductions and "unusual outages" experienced in Korea and Germany. Also, increases in the load factor for the current fleet of reactors appear to have plateaued.

The United States produces the most nuclear energy, with nuclear power providing 19 per cent of the electricity it consumes, while France produces the highest percentage of its electrical energy from nuclear reactors—78 per cent as of 2006. In the European Union as a whole, nuclear energy provides 30 per cent of the electricity. Nuclear energy policy differs between European Union countries, and some, such as

Austria and Ireland, have no active nuclear power stations. In comparison, France has a large number of these plants, with 16 multi-unit stations in current use.

In the US, while the Coal and Gas Electricity industry is projected to be worth $85 billion by 2013, Nuclear Power generators are forecast to be worth $18 billion.

Many military and some civilian (such as some icebreaker) ships use nuclear marine propulsion, a form of nuclear propulsion. A few space vehicles have been launched using full-fledged nuclear reactors: the Soviet RORSAT series and the American SNAP-10A.

International research is continuing into safety improvements such as passively safe plants, the use of nuclear fusion, and additional uses of process heat such as hydrogen production (in support of a hydrogen economy), for desalinating sea water, and for use in district heating systems.

Development of Nuclear Technology

Installed nuclear capacity initially rose relatively quickly, rising from less than 1 gigawatt (GW) in 1960 to 100 GW in the late 1970s, and 300 GW in the late 1980s. Since the late 1980s worldwide capacity has risen much more slowly, reaching 366 GW in 2005. Between around 1970 and 1990, more than 50 GW of capacity was under construction (peaking at over 150 GW in the late 70s and early 80s) — in 2005, around 25 GW of new capacity was planned.

More than two-thirds of all nuclear plants ordered after January 1970 were eventually cancelled. During the 1970s and 1980s rising economic costs (related to extended construction times largely due to regulatory changes and pressure-group litigation) and falling fossil fuel prices made nuclear power plants then under construction less attractive. In the 1980s (U.S.) and 1990s (Europe), flat load growth and electricity liberalization also made the addition of large new baseload capacity unattractive.

The 1973 oil crisis had a significant effect on countries, such as France and Japan, which had relied more heavily on oil for electric generation (39 per cent and 73 per cent respectively) to invest in nuclear power. Today, nuclear power supplies about 80 per cent and 30 per cent of the electricity in those countries, respectively. A general movement against nuclear power arose during the last third of the 20th century, based on the fear of a possible nuclear accident, fears of radiation, nuclear proliferation, and on the opposition to nuclear waste production, transport and final storage.

Perceived risks on the citizens' health and safety, the 1979 accident at Three Mile Island and the 1986 Chernobyl disaster played a part in stopping new plant construction in many countries, although the public policy organization Brookings Institution suggests that new nuclear units have not been ordered in the U.S. because the Institution's research concludes they cost 15–30 per cent more over their lifetime than conventional coal and natural gas fired plants.

Unlike the Three Mile Island accident, the much more serious Chernobyl accident did not increase regulations affecting Western reactors since the Chernobyl reactors were of the problematic RBMK design only used in the Soviet Union, for example lacking "robust" containment buildings. Many of these reactors are still in use today. However, changes were made in both the reactors themselves (use of low enriched uranium) and in the control system (prevention of disabling safety systems) to prevent the possibility of a duplicate accident.

An international organization to promote safety awareness and professional development on operators in nuclear facilities was created: WANO; World Association of Nuclear Operators. Opposition in Ireland, New Zealand and Poland prevented nuclear programs there, while Austria (1978), Sweden (1980) and Italy (1987) (influenced by Chernobyl) voted in referendums to oppose or phase out nuclear power.

Future of the Industry

As of 2007, Watts Bar 1, which came on-line in February 7, 1996, was the last U.S. commercial nuclear reactor to go on-line. This is often quoted as evidence of a successful worldwide campaign for nuclear power phase-out. However, political resistance to nuclear power has only ever been successful in New Zealand, parts of Europe, and the Philippines. Even in the U.S. and throughout Europe, investment in research and in the nuclear fuel cycle has continued, and some experts predict that electricity shortages, fossil fuel price increases, global warming and heavy metal emissions from fossil fuel use, new technology such as passively safe plants, and national energy security will renew the demand for nuclear power plants.

According to the World Nuclear Association, globally during the 1980s one new nuclear reactor started up every 17 days on average, and by the year 2015 this rate could increase to one every 5 days. Many countries remain active in developing nuclear power, including Pakistan, Japan, China and India, all actively developing both fast and thermal technology,

South Korea and the United States, developing thermal technology only, and South Africa and China, developing versions of the Pebble Bed Modular Reactor (PBMR). Several EU member states actively pursue nuclear programs, while some other member states continue to have a ban for the nuclear energy use. Japan has an active nuclear construction programme with new units brought on-line in 2005.

In the U.S., three consortia responded in 2004 to the U.S. Department of Energy's solicitation under the Nuclear Power 2010 Programme and were awarded matching funds—the Energy Policy Act of 2005 authorized loan guarantees for up to six new reactors, and authorized the Department of Energy to build a reactor based on the Generation IV Very-High-Temperature Reactor concept to produce both electricity and hydrogen.

As of the early 21st century, nuclear power is of particular interest to both China and India to serve their rapidly growing economies—both are developing fast breeder reactors. In the energy policy of the United Kingdom it is recognized that there is a likely future energy supply shortfall, which may have to be filled by either new nuclear plant construction or maintaining existing plants beyond their programmed lifetime.

There is a possible impediment to production of nuclear power plants, due to a backlog at Japan Steel Works, the only factory in the world able to manufacture the central part of a nuclear reactor's containment vessel in a single piece, which reduces the risk of a radiation leak. The company can only make four per year of the steel forgings. It will double its capacity in the next two years, but still will not be able to meet current global demand alone.

Utilities across the world are submitting orders years in advance of any actual need. Other manufacturers are examining various options, including making the component themselves, or finding ways to make a similar item using alternate methods. Other solutions include using designs that do not require single piece forged pressure vessles such as Canada's Advanced CANDU Reactors or Sodium-cooled Fast Reactors.

Other companies able to make the large forgings required for reactor pressure vessels include: Russia's OMZ, which is upgrading to be able to manufacture three or four pressure vessels per year; South Korea's Doosan Heavy Industries; and Mitsubishi Heavy Industries, which is doubling capacity for reactor pressure vessels and other large nuclear components. The UK's Sheffield Forgemasters is evaluating the benefit of tooling-up for nuclear forging work.

A 2007 status report from the anti-nuclear European Greens claimed that, "even if Finland and France build a European Pressurized water Reactor (EPR), China started an additional 20 plants and Japan, Korea or Eastern Europe added one or the other plant, the overall global trend for nuclear power capacity will most likely be downwards over the next two or three decades.

With extremely long lead times of 10 years and more [for plant construction], it is practically impossible to increase or even maintain the number of operating nuclear power plants over the next 20 years, unless operating lifetimes would be substantially increased beyond 40 years on average." In fact, China plans to build more than 100 plants, while in the US the licenses of almost half its reactors have already been extended to 60 years, and plans to build more than 30 new ones are under consideration.

In 2008, the International Atomic Energy Agency (IAEA) predicted that nuclear power capacity could double by 2030, though that would not be enough to increase nuclear's share of electricity generation.

NUCLEAR REACTOR TECHNOLOGY

Just as many conventional thermal power stations generate electricity by harnessing the thermal energy released from burning fossil fuels, nuclear power plants convert the energy released from the nucleus of an atom, typically via nuclear fission.

When a relatively large fissile atomic nucleus (usually uranium-235 or plutonium-239) absorbs a neutron, a fission of the atom often results. Fission splits the atom into two or more smaller nuclei with kinetic energy (known as fission products) and also releases gamma radiation and free neutrons. A portion of these neutrons may later be absorbed by other fissile atoms and create more fissions, which release more neutrons, and so on.

This nuclear chain reaction can be controlled by using neutron poisons and neutron moderators to change the portion of neutrons that will go on to cause more fissions. Nuclear reactors generally have automatic and manual systems to shut the fission reaction down if unsafe conditions are detected.

A cooling system removes heat from the reactor core and transports it to another area of the plant, where the thermal energy can be harnessed to produce electricity or to do other useful work. Typically the hot coolant will be used as a heat source for a boiler, and the pressurized steam from that boiler will power one or more steam turbine driven electrical generators.

There are many different reactors designs, utilizing different fuels and coolants and incorporating different control schemes. Some of these designs have been engineered to meet a specific need. Reactors for nuclear submarines and large naval ships, for example, commonly use highly enriched uranium as a fuel. This fuel choice increases the reactor's power density and extends the usable life of the nuclear fuel load, but is more expensive and a greater risk to nuclear proliferation than some of the other nuclear fuels. A number of new designs for nuclear power generation, collectively known as the Generation IV reactors, are the subject of active research and may be used for practical power generation in the future.

Many of these new designs specifically attempt to make fission reactors cleaner, safer and/or less of a risk to the proliferation of nuclear weapons. Passively safe plants (such as the ESBWR) are available to be built and other designs that are believed to be nearly fool-proof are being pursued. Fusion reactors, which may be viable in the future, diminish or eliminate many of the risks associated with nuclear fission.

NUCLEAR FUEL CYCLE

Fig. Nuclear Fuel Cycle

The Nuclear Fuel Cycle begins when uranium is mined, enriched, and manufactured into nuclear fuel, (1) which is delivered to a nuclear power plant. After usage in the power plant, the spent fuel is delivered to a reprocessing plant (2) or to a final repository (3) for geological disposition. In reprocessing 95 per cent of spent fuel can be recycled to be returned to usage in a power plant (4).

A nuclear reactor is only part of the life-cycle for nuclear power. The process starts with mining. Uranium mines are underground, open-pit, or in-situ leach mines. In any case, the uranium ore is extracted, usually converted into a stable and compact form such as yellowcake, and then transported to a processing facility. Here, the yellowcake is converted to uranium hexafluoride, which is then enriched using various techniques.

At this point, the enriched uranium, containing more than the natural 0.7 per cent U-235, is used to make rods of the proper composition and geometry for the particular reactor that the fuel is destined for. The fuel rods will spend about 3 operational cycles (typically 6 years total now) inside the reactor, generally until about 3 per cent of their uranium has been fissioned, then they will be moved to a spent fuel pool where the short lived isotopes generated by fission can decay away. After about 5 years in a cooling pond, the spent fuel is radioactively and thermally cool enough to handle, and it can be moved to dry storage casks or reprocessed.

CONVENTIONAL FUEL RESOURCES

Uranium is a fairly common element in the Earth's crust. Uranium is approximately as common as tin or germanium in Earth's crust, and is about 35 times more common than silver. Uranium is a constituent of most rocks, dirt, and of the oceans.

The world's present measured resources of uranium, economically recoverable at a price of 130 USD/kg, are enough to last for "at least a century" at current consumption rates. This represents a higher level of assured resources than is normal for most minerals. On the basis of analogies with other metallic minerals, a doubling of price from present levels could be expected to create about a tenfold increase in measured resources, over time. The fuel's contribution to the overall cost of the electricity produced is relatively small, so even a large fuel price escalation will have relatively little effect on final price.

For instance, typically a doubling of the uranium market price would increase the fuel cost for a light water reactor by 26 per cent and the electricity cost about 7 per cent, whereas doubling the price of natural gas would typically add 70 per cent to the price of electricity from that source.

At high enough prices, eventually extraction from sources such as granite and seawater become economically feasible. Current light water reactors make relatively inefficient use of nuclear fuel, fissioning only the very rare uranium-235 isotope. Nuclear reprocessing can make this waste reusable and more efficient reactor designs allow better use of the available resources.

BREEDING

As opposed to current light water reactors which use uranium-235 (0.7 per cent of all natural uranium), fast breeder reactors use uranium-238 (99.3 per cent of all natural uranium). It has been estimated that there is up to five billion years' worth of uranium-238 for use in these power plants. Breeder technology has been used in several reactors, but the high cost of reprocessing fuel safely requires uranium prices of more than 200 USD/kg before becoming justified economically.

As of December 2005, the only breeder reactor producing power is BN-600 in Beloyarsk, Russia. The electricity output of BN-600 is 600 MW — Russia has planned to build another unit, BN-800, at Beloyarsk nuclear power plant. Also, Japan's Monju reactor is planned for restart (having been shut down since 1995), and both China and India intend to build breeder reactors.

Another alternative would be to use uranium-233 bred from thorium as fission fuel in the thorium fuel cycle. Thorium is about 3.5 times as common as uranium in the Earth's crust, and has different geographic characteristics. This would extend the total practical fissionable resource base by 450 per cent.

Unlike the breeding of U-238 into plutonium, fast breeder reactors are not necessary — it can be performed satisfactorily in more conventional plants. India has looked into this technology, as it has abundant thorium reserves but little uranium.

FUSION

Fusion power advocates commonly propose the use of deuterium, an isotope of hydrogen, as fuel and in many current designs also lithium. Assuming a fusion energy output equal to the current global output and that this does not increase in the future, then the known current lithium reserves would last 3000 years, lithium from sea water would last 60 million years, and a more complicated fusion process using only deuterium from sea water would have fuel for 150 billion years.

WATER

Like all forms of power generation using steam turbines, Nuclear power plants use large amounts of water for cooling. At Sellafield, which is no longer producing electricity, a maximum of 18,184.4 m^3 a day (over 4 million gallons) and 6,637,306 m^3 a year of fresh water from Wast Water is still abstracted to use on site for various processes. As with most power plants, two-thirds of the energy produced by a nuclear power plant goes into waste heat, and that heat is carried away from the plant in the water (which remains uncontaminated by radioactivity).

The emitted water either is sent into cooling towers where it goes up and is emitted as water droplets (literally a cloud) or is discharged into large bodies of water — cooling ponds, lakes, rivers, or oceans. Droughts can pose a severe problem by causing the source of cooling water to run out.

The Palo Verde Nuclear Generating Station near Phoenix, AZ is the only nuclear generating facility in the world that is not located adjacent to a large body of water. Instead, it uses treated sewage from several nearby municipalities to meet its cooling water needs, recycling 20 billion US gallons (76,000,000 m^3) of wastewater each year. Like conventional power plants, nuclear power plants generate large quantities of waste heat which is expelled in the condenser, following the turbine.

Colocation of plants that can take advantage of this thermal energy has been suggested by Oak Ridge National Laboratory (ORNL) as a way to take advantage of process synergy for added energy efficiency. One example would be to use the power plant steam to produce hydrogen from water. The hydrogen would cost less, and the nuclear power plant would exhaust less heat into the atmosphere and water vapour, which is a short-lived greenhouse gas.

SOLID WASTE

The safe storage and disposal of nuclear waste is a significant challenge. The most important waste stream from nuclear power plants is spent fuel. A large nuclear reactor produces 3 cubic metres (25–30 tonnes) of spent fuel each year. It is primarily composed of unconverted uranium as well as significant quantities of transuranic actinides (plutonium and curium, mostly).

In addition, about 3 per cent of it is made of fission products. The actinides (uranium, plutonium, and curium) are responsible for the bulk of the long term radioactivity, whereas the fission products are responsible for the bulk of the short term radioactivity.

HIGH LEVEL RADIOACTIVE WASTE

Spent fuel is highly radioactive and needs to be handled with great care and forethought. However, spent nuclear fuel becomes less radioactive over time. After 40 years, the radiation flux is 99.9 per cent lower than it was the moment the spent fuel was removed, although still dangerously radioactive.

Spent fuel rods are stored in shielded basins of water (spent fuel pools), usually located on-site. The water provides both cooling for the still-decaying fission products, and shielding from the continuing radioactivity. After a few decades some on-site storage involves moving the now cooler, less radioactive fuel to a dry-storage facility or dry cask storage, where the fuel is stored in steel and concrete containers until its radioactivity decreases naturally ("decays") to levels safe enough for other processing. This interim stage spans years or decades, depending on the type of fuel. Most U.S. waste is currently stored in temporary storage sites requiring oversight, while suitable permanent disposal methods are discussed.

As of 2007, the United States had accumulated more than 50,000 metric tons of spent nuclear fuel from nuclear reactors. Underground storage at Yucca Mountain in U.S. has been proposed as permanent storage.

After 10,000 years of radioactive decay, according to United States Environmental Protection Agency standards, the spent nuclear fuel will no longer pose a threat to public health and safety.

The amount of waste can be reduced in several ways, particularly reprocessing. Even so, the remaining waste will be substantially radioactive for at least 300 years even if the actinides are removed, and for up to thousands of years if the actinides are left in.

Even with separation of all actinides, and using fast breeder reactors to destroy by transmutation some of the longer-lived non-actinides as well, the waste must be segregated from the environment for one to a few hundred years, and therefore this is properly categorized as a long-term problem.

Subcritical reactors or fusion reactors could also reduce the time the waste has to be stored. It has been argued that the best solution for the nuclear waste is above ground temporary storage since technology is rapidly changing. The current waste may well become a valuable resource in the future.

LOW-LEVEL RADIOACTIVE WASTE

The nuclear industry also produces a volume of low-level radioactive waste in the form of contaminated items like clothing, hand tools, water purifier resins, and (upon decommissioning) the materials of which the reactor itself is built. In the United States, the Nuclear Regulatory Commission has repeatedly attempted to allow low-level materials to be handled as normal waste: landfilled, recycled into consumer items, et cetera. Most low-level waste releases very low levels of radioactivity and is only considered radioactive waste because of its history. For example, according to the standards of the NRC, the radiation released by coffee is enough to treat it as low level waste.

COMPARING RADIOACTIVE WASTE TO INDUSTRIAL TOXIC WASTE

In countries with nuclear power, radioactive wastes comprise less than 1 per cent of total industrial toxic wastes, which remain hazardous indefinitely unless they decompose or are treated so that they are less toxic or, ideally, completely non-toxic. Overall, nuclear power produces far less waste material than fossil-fuel based power plants.

Coal-burning plants are particularly noted for producing large amounts of toxic and mildly radioactive ash due to concentrating naturally occurring metals and radioactive material from the coal. Contrary to popular belief, coal power actually results in more radioactive waste being released into the environment than nuclear power. The population effective dose equivalent from radiation from coal plants is 100 times as much as nuclear plants.

REPROCESSING

Reprocessing can potentially recover up to 95 per cent of the remaining uranium and plutonium in spent nuclear fuel, putting it into new mixed oxide fuel. This produces a reduction in long term radioactivity within the remaining waste, since this is largely short-lived fission products, and reduces its volume by over 90 per cent.

Reprocessing of civilian fuel from power reactors is currently done on large scale in Britain, France and (formerly) Russia, soon will be done in China and perhaps India, and is being done on an expanding scale in Japan. The full potential of reprocessing has not been achieved because it requires breeder reactors, which are not yet commercially available. France is generally cited as the most successful reprocessor, but it presently only recycles 28 per cent (by mass) of the yearly fuel use, 7 per cent within France and another 21 per cent in Russia.

Unlike other countries, the US stopped civilian reprocessing from 1976 to 1981 as one part of US non-proliferation policy, since reprocessed material such as plutonium could be used in nuclear weapons: however, reprocessing is now allowed in the U.S. Even so, in the U.S. spent nuclear fuel is currently all treated as waste.

In February, 2006, a new U.S. initiative, the Global Nuclear Energy Partnership was announced. It would be an international effort to reprocess fuel in a manner making nuclear proliferation unfeasible, while making nuclear power available to developing countries.

DEPLETED URANIUM

Uranium enrichment produces many tons of depleted uranium (DU) which consists of U-238 with most of the easily fissile U-235 isotope removed. U-238 is a tough metal with several commercial uses — for example, aircraft production, radiation shielding, and armor — as it has a higher density than lead. Depleted uranium is also useful in munitions as DU penetrators (bullets or APFSDS tips) 'self sharpen', due to uranium's tendency to fracture along adiabatic shear bands.

There are concerns that U-238 may lead to health problems in groups exposed to this material excessively, like tank crews and civilians living in areas where large quantities of DU ammunition have been used.

In January 2003 the World Health Organization released a report finding that contamination from DU munitions were localized to a few tens of meters from the impact sites and contamination of local vegetation and water was 'extremely low'. The report also states that approximately 70 per cent of ingested DU will leave the body after twenty four hours and 90 per cent after a few days.

Chapter 7

Radiography

Radiography is the use of X-rays to view unseen or hard-to-image objects. The main diagnostic purposes of X-rays are to see inside ones' body, especially of the brain and fetus, where the bones can be viewed at an optimum resolution. The impact on society of this technique has also been immense. Physicists and researchers have developed numerous types of medicines to assist in the function and development of organs such as the brain and heart, and radiography has improved the economy of most countries in the western world due to the employment of physicists and doctors.

Radiography is used for both medical and industrial applications. If the object being examined is living, whether human or animal, it is regarded as medical; all other radiography is regarded as industrial radiographic work.

EQUIPMENT

A number of sources of X-ray photons have been used; these include sealed X-ray tubes, betatrons, and linear accelerators (linacs). For gamma rays, radioactive sources such as ^{192}Ir have been used.

A range of detectors including photographic film, scintillator and semiconductor diode arrays have been used to collect images.

Theory of X-ray Attenuation

X-ray photons used for medical purposes are formed by an event involving an electron, while gamma ray photons are formed from an interaction with the nucleus of an atom. In general, medical radiography is done using X-rays formed in an X-ray tube. Nuclear medicine typically involves gamma rays.

The types of electromagnetic radiation of most interest to radiography are X-ray and gamma radiation. This radiation is much more energetic than the more familiar types such as radio waves and visible light. It is this relatively high energy which makes gamma rays useful in radiography but potentially hazardous to living organisms.

The radiation is produced by X-ray tubes, high energy X-ray equipment or natural radioactive elements, such as radium and radon, and artificially produced radioactive isotopes of elements, such as cobalt-60 and iridium-192. Electromagnetic radiation consists of oscillating electric and magnetic fields, but is generally depicted as a single sinusoidal wave. While in the past radium and radon have both been used for radiography, they have fallen out of use as they are radiotoxic alpha radiation emitters which are expensive; iridium-192 and cobalt-60 are far better photon sources.

Such a wave is characterised by its wavelength (the distance from a point on one cycle to the corresponding point on the next cycle) or its frequency (the number of oscillations per second). In a vacuum, all electromagnetic waves travel at the same speed, the speed of light (*c*). The wavelength (ë, lambda) and the frequency (f) are all related by the equation:

$$f = \frac{c}{\lambda}$$

This is true for all electromagnetic radiation.

Electromagnetic radiation is known by various names, depending on its energy. The energy of these waves is related to the frequency and the wavelength by the relationship:

$$E = hf = h\left(\frac{c}{\lambda}\right)$$

where *h* is a constant known as the Planck constant.

Gamma rays are indirectly ionizing radiation. A gamma ray passes through matter until it undergoes an interaction with an atomic particle, usually an electron. During this interaction, energy is transferred from the gamma ray to the electron, which is a directly ionizing particle. As a result of this energy transfer, the electron is liberated from the atom and proceeds to ionize matter by colliding with other electrons along its path. Other times, the passing gamma ray interferes with the orbit of the electron, and slows it, releasing energy but not becoming dislodged.

The atom is not ionised, and the gamma ray continues on, although at a lower energy. This energy released is usually heat or another, weaker photon, and causes biological harm as a radiation burn. The chain reaction caused by the initial dose of radiation can continue after exposure, much like a sunburn continues to damage skin even after one is out of direct sunlight.

For the range of energies commonly used in radiography, the interaction between gamma rays and electrons occurs in two ways. One effect takes place where all the gamma ray's energy is transmitted to an entire atom. The gamma ray no longer exists and an electron emerges from the atom with kinetic (motion in relation to force) energy almost equal to the gamma energy.

This effect is predominant at low gamma energies and is known as the photoelectric effect. The other major effect occurs when a gamma ray interacts with an atomic electron, freeing it from the atom and imparting to it only a fraction of the gamma ray's kinetic energy. A secondary gamma ray with less energy (hence lower frequency) also emerges from the interaction. This effect predominates at higher gamma energies and is known as the Compton effect.

In both of these effects the emergent electrons lose their kinetic energy by ionizing surrounding atoms. The density of ions so generated is a measure of the energy delivered to the material by the gamma rays.

The most common means of measuring the variations in a beam of radiation is by observing its effect on a photographic film. This effect is the same as that of light, and the more intense the radiation is, the more it darkens, or exposes, the film. Other methods are in use, such as the ionizing effect measured electronically, its ability to discharge an electrostatically charged plate or to cause certain chemicals to fluoresce as in fluoroscopy.

MEDICAL RADIOGRAPHY

Radiography is the use of ionizing electromagnetic radiation such as X-rays to view objects. Although not technically radiographic techniques, imaging modalities such as PET and MRI are sometimes grouped in radiography due to the fact that the radiology department of hospitals handle all forms of imaging. Treatment using radiation is known as radiotherapy.

DIAGNOSTIC RADIOGRAPHY

Diagnostic radiography involves the use of both ionising radiation and non-ionising radiation to create images for medical diagnoses. The predominant test is still the X-ray (the word X-ray is often used for both the test and the actual film or digital image). X-rays are the second most commonly used medical tests, after laboratory tests. This application is known as diagnostic radiography.

Since the body is made up of various substances with differing densities, X-rays can be used to reveal the internal structure of the body on film by highlighting these differences using attenuation, or the *absorption* of X-ray photons by the denser substances (like calcium-rich bones). Medical diagnostic radiography is undertaken by a specially trained professional called a diagnostic radiographer in the UK, or a radiologic technologist in the USA.

PROJECTION RADIOGRAPHY

The creation of images by exposing an object to X-rays or other high-energy forms of electromagnetic radiation and capturing the resulting remnant beam (or "shadow") as a latent image is known as "projection radiography." The "shadow" may be converted to light using a fluorescent screen, which is then captured on photographic film, it may be captured by a phosphor screen to be "read" later by a laser (CR), or it may directly activate a matrix of solid-state detectors (DR—similar to a very large version of a CCD in a digital camera). Bone and some organs (such as lungs) especially lend themselves to projection radiography. It is a relatively low-cost investigation with a high diagnostic yield.

Projection radiography uses X-rays in different amounts and strengths depending on what body part is being imaged:

- Hard tissues such as bone require a relatively high energy photon source, and typically a tungsten anode is used with a high voltage (50-150 kVp) on a 3-phase or high-frequency machine to generate braking radiation. Bony tissue and metals are denser than the surrounding tissue, and thus by absorbing more of the X-ray photons they prevent the film from getting exposed as much.

  Wherever dense tissue absorbs or stops the X-rays, the resulting X-ray film is unexposed, and appears translucent blue, whereas the black parts of the film represent lower-density tissues such as fat, skin, and internal organs, which could not stop the X-rays. This is usually used to see bony fractures, foreign objects (such as ingested coins), and used for finding bony pathology such as osteoarthritis, infection (osteomyelitis), cancer (osteosarcoma), as well as growth studies (leg length, achondroplasia, scoliosis, etc).
- Soft tissues are seen with the same machine as for hard tissues, but a "softer" or less-penetrating X-ray beam is used. Tissues commonly imaged include the lungs and heart shadow in a chest X-ray, the air pattern of the bowel in abdominal X-rays, the soft tissues of the neck, the orbits by a skull X-ray before an MRI to check for radiopaque foreign bodies (especially metal), and of course the soft tissue shadows in X-rays of bony injuries are looked at by the radiologist for signs of hidden trauma (for example, the famous "fat pad" sign on a fractured elbow).

- Dental radiography uses a small radiation dose with high penetration to view teeth, which are relatively dense. A dentist may examine a painful tooth and gum using X-ray equipment. The machines used are typically single-phase pulsating DC, the oldest and simplest sort. Dental technicians or the dentist may run these machines— radio logic technologists are not required by law to be present.
- Mammography is an X-ray examination of breasts and other soft tissues. This has been used mostly on women to screen for breast cancer, but is also used to view male breasts, and used in conjunction with a radiologist or a surgeon to localise suspicious tissues before a biopsy or a lumpectomy.

  Breast implants designed to enlarge the breasts reduce the viewing ability of mammography, and require more time for imaging as more views need to be taken.

  This is because the material used in the implant is very dense compared to breast tissue, and looks white (clear) on the film. The radiation used for mammography tends to be softer (has a lower photon energy) than that used for the harder tissues. Often a tube with a molybdenum anode is used with about 30 000 volts (30 kV), giving a range of X-ray energies of about 15-30 keV. Many of these photons are "characteristic radiation" of a specific energy determined by the atomic structure of the target material (Mo-K radiation).

Fluoroscopy

Fluoroscopy is a term invented by Thomas Edison during his early X-ray studies. The name refers to the fluorescence he saw while looking at a glowing plate bombarded with X-rays.

This is a technique that provides moving projection radiographs of lower quality. Fluoroscopy is mainly performed to view movement (of tissue or a contrast agent), or to guide a medical intervention, such as angioplasty, pacemaker insertion, or joint repair/replacement. The latter are often carried out in the operating theatre, using a portable fluoroscopy machine called a C-arm. It can move around the surgery table and make digital images for the surgeon.

Angiography is the use of fluoroscopy to view the cardiovascular system. An iodine-based contrast is injected into the bloodstream and watched as it travels around. Since liquid blood and the vessels are not very dense, a contrast with high density (like the large iodine atoms) is used to view the vessels under X-ray. Angiography is used to find aneurysms, leaks, blockages (thromboses), new vessel growth, and placement of catheters and stents. Balloon angioplasty is often done with angiography.

Fluoroscopy can be used to examine the digestive system using a substance which is opaque to X-rays, (usually barium sulfate or gastrografin), which is introduced into the digestive system either by swallowing or as an enema. This is normally as part of a *double contrast technique,* using positive and negative contrast. Barium sulfate coats the walls of the digestive tract (positive contrast), which allows the shape of the digestive tract to be outlined as white or clear on an X-ray.

Air may then be introduced (negative contrast), which looks black on the film. The barium meal is an example of a contrast agent swallowed to examine the upper digestive tract. Note that while soluble barium compounds are very toxic, the insoluble barium sulfate is non-toxic because its low solubility prevents the body from absorbing it.

- A number of substances have been used as positive contrast agents: silver, bismuth, caesium, thorium, tin, zirconium, tantalum, tungsten and lanthanide compounds have been used as contrast agents. The use of thoria (thorium dioxide) as an agent was rapidly stopped as thorium causes liver cancer.

Most modern injected radiographic positive contrast media are iodine-based. Patients who suffer from allergy to shellfish may be allergic to iodine, and should consult their physician regarding pre-medication to lessen risk of allergic reaction. Iodinated contrast comes in two forms: ionic and non-ionic compounds. Non-ionic contrast is significantly more expensive than ionic (approximately three to five times the cost), however, non-ionic contrast tends to be safer for the patient, causing fewer allergic reactions and uncomfortable side effects such as hot sensations or flushing. Most imaging centers now use non-ionic contrast exclusively, finding that the benefits to patients outweigh the expense.

- Negative radiographic contrast agents are air and carbon dioxide (CO_2). The latter is easily absorbed by the body and causes less spasm. It can also be injected into the blood, where air absolutely cannot.

Dual energy X-ray Absorptiometry

DEXA, or bone densitometry, is used primarily for osteoporosis tests. It is not projection radiography, as the X-rays are emitted in 2 narrow beams that are scanned across the patient, 90 degrees from each other. Usually the hip (head of the femur), lower back (lumbar spine) or heel (calcaneum) are imaged, and the bone density (amount of calcium) is determined and given a number (a T-score).

It is not used for bone imaging, as the image quality is not good enough to make an accurate diagnostic image for fractures, inflammation etc. It can also be used to measure total body fat, though this isn't common. The radiation dose received from DEXA scans is very low, much lower than projection radiography examinations.

Computed Tomography

Computed tomography or CT scan (previously known as CAT scan, the "A" standing for "axial") uses a high amount of ionizing radiation (in the form of X-rays) in conjunction with a computer to create images of both soft and hard tissues. These images look as though the patient was sliced like bread (thus, "tomography"— "tomo" means "slice"). The machine looks similar to an MRI machine to many patients, but is not related.

The exams are generally short, most lasting only as long as a breath-hold. Contrast agents are often used, depending on the tissues needing to be seen. Radiographers perform these examinations, sometimes in conjunction with a radiologist (for instance, when a radiologist performs a CT-guided biopsy).

TECHNICAL CONSIDERATIONS

X-ray photons are formed in events involving electrons and are the mainly form of ionizing electromagnetic radiation used in medical radiography. This radiation is much more energetic than the more familiar types such as radio waves and visible light. Proper production and detection of photons are important in the creation of good radiograms.

Photon Production

X-ray radiation for medical imaging is typically produced by X-ray tubes, which operate through bombarding the anode with high energy electrons emitted from a hot cathode. Image sharpness, contrast, and patient dosage are important considerations in medical radiography and these requirements determined the desired energies of the tube, the type of material used on the anode, and the method in which the power is generated to drive the tube. Although the technical definition of x-rays rage from 1-700 keV, medical x-rays use typically of 5-150 keV x-rays.

The photons emitted come in discrete bands of energy corresponding to the material of the anode, and the undesired band are removed. Choice of the anode and its emitted radiation energies depends on the application and the tissues being imaged, for instance molybdenum is often used in mammography due to it 20 keV x-rays. Too high radiation energies will result in poor pictures since the radiation cannot be readily attenuated, however too low energies will increase the radiation dosage of the patient without improvements in image quality.

Sharpness of a radiographic image is strongly determined by the the size of the x-ray source. This is determined by the are of the electron beam hitting the anode. A large photon sources results in more blurring in the final image and is worsened a increase in image formation distance. This blurring can be measured as a contribution to the modulation transfer function of the imaging system.

Power Generation

The power used by the x-ray tube is generated by a specialized generator, which supplies the voltage and current required to drive the tube. The generator needs to supply high voltages with small exposure times. An exposure thus can be described by two factor:

- The *peak voltage* of the cathode to anode
- The *milliamprere seconds* exposure time

These variables can be controlled by the operator but is more typically assigned automatically by the x-ray machinery through sampling the emitted radiation. Power generators convert standard 120 or 220 volt AC to higher DC voltages and typically employ rectified and filtered multiphase transformers which maintain a constant voltage and can be turn rapidly on and off for millisecond exposures.

Photon Detection

Photons images that have been shadowed from an imaging subject must be detected at high fidelity and resolution to allow for diagnosis. There are three main types of image detection methods used namely: film/screens, image intensifiers, and digital detectors, with the latter fast becoming the standard for x-ray image detection.

Film/Screens

X-ray film is almost always used in conjunction with x-ray sensitive screen due to the fact that high resolution film is quite poor at detecting x-rays. These screens contain rare earth minerals and phosphor materials that convert x-ray radiation to visible light or lows EM energies to which the film is sensitive. Screen generally have to have good contrast, dynamic range, and resolution, with the former two factors being competing properties.

The resolution of the screen also affects the sensitivity of the detectors since more sensitive screens are generally thicken, which causes the more blurring due to spreading light. The film speed also play a factor in image quality. Higher speeds are more sensitive to photons but are generally lower in resolution and more susceptible to noise. Lower speed films produce images of good resolution and dynamic range but requires more photons for exposure and increases the radiation dosage of the subject.

Image Intensifiers and Array Detectors

Image intensifiers are analog devices that readily convert the acquired x-ray image into one visible on an video screen. This device is made of a vacuum tube with a wide input surface coated on the inside with caesium iodide (CsI). When hit by x-rays material phosphors which causes the photocathode adjacent to it to emit electrons. These electron are then focus using electron lenses inside the intensifier to a output screen coated with phosphorescent materials. The image from the output can then be recorded via a camera and displayed.

Digital devices known an array detectors are becoming more common in fluoroscopy. These devices are made of discrete pixelated detectors known as TFTs which can either work *indirectly* by using photo detectors that detect light emitted from a scintillator material such as CsI, or *directly* by capturing the electrons produced when the x-rays hit the detector. Direct detector do not tend to experience the blurring or spreading effect caused by phosphorescent scintillators of or film screens since the detectors are activated directly by x-ray photons.

INDUSTRIAL RADIOGRAPHY

Industrial Radiography is the use of ionising electromagnetic radiation to view objects in a way that can't be seen otherwise. It is not to be confused with the use of ionizing radiation to change or modify objects; radiography's purpose is strictly for viewing. Industrial radiography has grown out of engineering, and is a major element of nondestructive testing (NDT). It is a method of inspecting materials for hidden flaws by using the ability of short wavelength electromagnetic radiation to penetrate various materials.

USES OF RADIOGRAPHY IN INDUSTRIES

Inspection of Materials

Gamma radiation sources, most commonly Iridium-192 and Cobalt-60, are used to inspect a variety of materials. The vast majority of radiography concerns the testing and grading of welds on pressurized piping, pressure vessels, high-capacity storage containers, pipelines, and some structural welds.

Other tested materials include concrete (locating rebar or conduit), welder's test coupons, machined parts, plate metal, or pipewall (locating anomalies due to corrosion or mechanical damage). Theoretically, industrial radiographers could radiograph any solid, flat material (walls, ceilings, floors, square or rectangular containers) or any hollow cylindrical or spherical object. For purposes of inspection, including weld inspection, there exist several exposure arrangements.

First, there is the panoramic, one of the four single wall exposure/single wall view (SWE/SWV) arrangements. This exposure is created when the radiographer places the source of radiation at the centre of a sphere, cone, or cylinder (including tanks, vessels, and piping). Depending upon client requirements, the radiographer would then place film cassettes on the outside of the surface to be examined. This exposure arrangement is ideal - when properly arranged and exposed, all portions of all exposed film will be of the same approximate density.

It also has the advantage of taking less time than other arrangements since the source must only penetrate the total wall thickness (WT) once and must only travel the radius of the inspection item, not its full diameter. The major disadvantage of the panoramic is that it may be impractical to reach the centre of the item (enclosed pipe) or the source may be too weak to perform in this arrangement (large vessels or tanks).

The second SWE/SWV arrangement is an interior placement of the source in an enclosed inspection item without having the source centered up. The source does not come in direct contact with the item, but is placed a distance away, depending on client requirements. The third is an exterior placement with similar characteristics.

The fifth is reserved for flat objects, such as plate metal, and is also radiographed without the source coming in direct contact with the item. In each case, the radiographic film is located on the opposite side of the inspection item from the source. In all four cases, only one wall is exposed, and only one wall is viewed on the radiograph.

Of the other exposure arrangements, only the contact shot has the source located on the inspection item. This type of radiograph will expose both walls, but will only resolve the image of that wall which is nearest the film.

This exposure arrangement takes more time than a panoramic, as the source must now penetrate the WT twice and travel the entire outside diameter of the pipe or vessel to reach the film on the opposite side.

This is a double wall exposure/single wall view DWE/SWV arrangement. Another is the superimposure (wherein the source is placed on one side of the item, not in direct contact with it, with the film on the opposite side).

This arrangement is usually reserved for very small diameter piping or parts. The last DWE/SWV exposure arrangement is the elliptical, in which the source is offset from the plane of the inspection item (usually a weld in pipe) and the elliptical image of the weld furthest from the source is cast onto the film.

The beam of radiation must be directed to the middle of the section under examination and must be normal to the material surface at that point, except in special techniques where known defects are best revealed by a different alignment of the beam.

The length of weld under examination for each exposure shall be such that the thickness of the material at the diagnostic extremities, measured in the direction of the incident beam, does not exceed the actual thickness at that point by more than 6 per cent. The specimen to be inspected is placed between the source of radiation and the detecting device, usually the film in a light tight holder or cassette, and the radiation is allowed to penetrate the part for the required length of time to be adequately recorded.

The result is a two-dimensional projection of the part onto the film, producing a latent image of varying densities according to the amount of radiation reaching each area. It is known as a radiograph, as distinct from a photograph produced by light.

Because film is cumulative in its response (the exposure increasing as it absorbs more radiation), relatively weak radiation can be detected by prolonging the exposure until the film can record an image that will be visible after development. The radiograph is examined as a negative, without printing as a positive as in photography. This is because, in printing, some of the detail is always lost and no useful purpose is served.

Before commencing a radiographic examination, it is always advisable to examine the component with one's own eyes, to eliminate any possible external defects. If the surface of a weld is too irregular, it may be desirable to grind it to obtain a smooth finish, but this is likely to be limited to those cases in which the surface irregularities (which will be visible on the radiograph) may make detecting internal defects difficult.

After this visual examination, the operator will have a clear idea of the possibilities of access to the two faces of the weld, which is important both for the setting up of the equipment and for the choice of the most appropriate technique.

Airport Security

Both hold luggage and carry-on hand luggage are normally examined by X-ray machines using X-ray radiography. Non-intrusive Cargo Scanning (aka Non-Intrusive Inspection - NII)

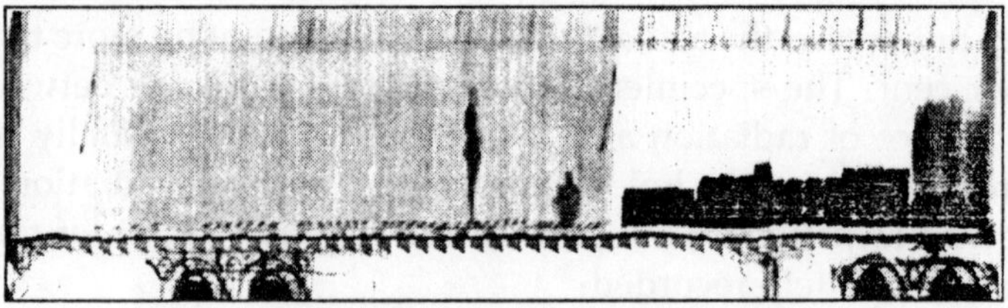

Fig. Gamma-ray Image of Intermodal Cargo Container with Stowaways

Gamma Radiography and High-Energy X-ray radiography are currently used to scan intermodal freight cargo containers in US and other countries. Also research is being done on adapting other types of radiography like Dual-Energy X-ray Radiography or Muon Radiography for scanning intermodal cargo containers.

SOURCES OF RADIOGRAPHY

X-ray Sources

A high energy X-ray machine can be used. It is often important to use a high accelerating voltage to provide the electrons with a very high energy. This is because in a braking radiation source the maximum photon energy is determined by the energy of the charged particles. A recent development is the betatron, which is a device similar to a cyclotron that acts as a very intense photon source.

Radioisotope Sources

These have the advantage that they do not need a supply of electrical power to function, but they do have the disadvantage that they can not be turned off. Also it is difficult using radioactivity to create a small and compact source which offers the photon flux which is possible with a normal sealed X-ray tube.

It might be possible to use Cs-137 as a photon source for radiography but this isotope is always diluted with inactive caesium isotopes. This makes it difficult to get a physcially small source, and a large volume of the source makes it impossible to capture fine details in a radiographic examination. Both cobalt-60 and caesium-137 have only a few gamma energies which make them close to monochromatic. The photon energy of cobalt-60 is higher than that of caesium-137, which allows cobalt sources to be used to examine thicker sections of metals than those which could be examined with Cs-137. Iridium-192 has a lower photon energy than cobalt-60 and its gamma spectrum is complex (many lines of very different enegies), but this can be an advantage as this can give better contrast for the final photographs.

It has been known for many years that an inactive iridium or cobalt metal object can be machined to size. In the case of cobalt it is common to alloy it with nickel to improve the mechanical properties. In the case of iridium a thin wire or rod could be used. These precursor materials can then be placed within stainless steel containers which are leak tested before being converted into radioactive sources. These objects can be processed by neutron activation to form gamma emitting radioisotopes.

The stainless steel has only a small ability to be activated and the small activity due to ^{55}Fe and ^{63}Ni are unlikely to pose a problem in the final application because these isotopes are beta emitters which have very weak gamma emission. The ^{59}Fe which might form has a short half life, so by allowing a cobalt source to stand for a year much of this isotope will decay away.

The source is often a very small object which needs to be transported to the site where the work is to be conducted in a shielded container. It is normal to place the film in industrial radiography, clear the area where the work is to be done, add shielding (collimators) to reduce the size of the controlled area before exposing the radioactive source.

A series of different designs have been developed for radiographic "cameras". Rather than the "camera" being a device which accepts photons to record a picture, the "camera" in industrial radiography is the radioactive photon source.

Torch Design of Radiographic Cameras

One design is best thought of as being like a torch. The radioactive source is placed inside a shielded box, a hinge allows part of the shielding to be opened exposing the source, allowing photons to exit the radiography camera.

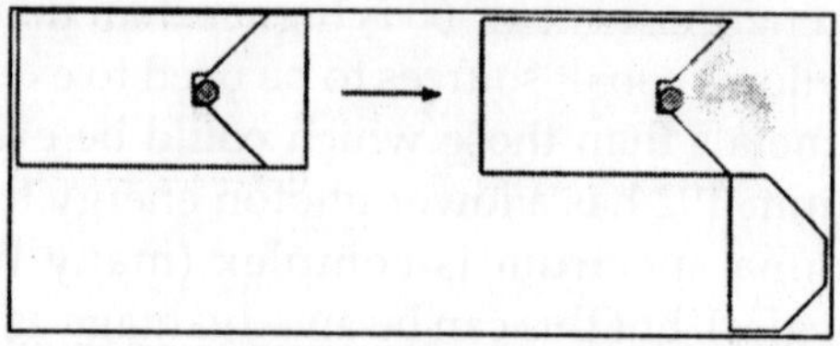

Fig. The Radiography Camera

A torch type camera which uses a hinge. The radioactive source is in red, the shielding by blue/green and the gamma rays are represented by yellow. Another design for a torch is where the source is placed in a metal wheel, which can turn inside the camera to move between the expose and storage positions.

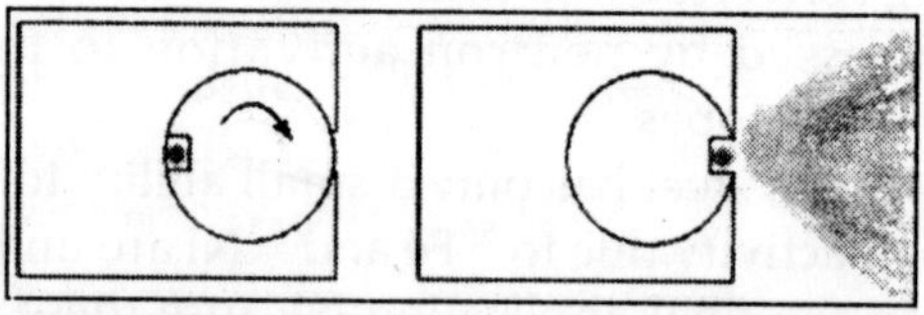

Fig. A Torch Type Camera which Uses a Wheel Design.

The radioactive source is in red and the gamma rays are represented by yellow

Cable Based Design of Radiographic Cameras

One group of designs use a radioactive source which comes out on a cable from a shielded container. One such unit was involved in an accident which occurred in Bolivia. This type of radiography could be compared to the remote afterloading method in Brachytherapy. In one design of equipment the source is stored in a block of lead or DU metal which has a S shaped tube like hole which passes through the block.

In the safe position the source is in the centre of the block and is attached to a metal wire which extends in both directions, to use the source a guide tube is attached to one side of the block while a drive cable is attached to the other end of the short cable. Using a hand operated winch the source is then pushed out of the shield and along the guide tube to were it is needed. It was one of these cable-based systems which was involved in an accident which occurred in Bolivia.

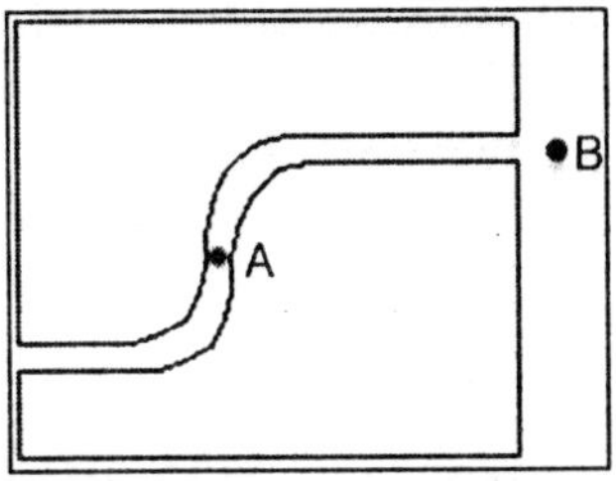

Fig. A Diagram of the Shaped Hole which Passes Through a Metal Block.

The source is stored at point A and it is driven out on a cable through a hole to point B. It will often go a long way along a guide tube to where it is needed.

Contrast Agents

Defects such as delaminations and planar cracks are difficult to detect using radiography, which is why penetrants are often used to enhance the contrast in the detection of such defects. Penetrants used include silver nitrate, zinc iodide, chloroform and diiodomethane.

Choice of the penetrant is determined by the ease with which it can penetrate the cracks and also with which it can be removed. Diiodomethane has the advantages of high opacity, ease of penetration, and ease of removal because it evaporates relatively quickly. However, it can cause skin burns.

Neutrons

In some rare cases, radiography is done with neutrons. This type of radiography is called *neutron radiography* (NR, Nray, N-Ray) or *neutron imaging*. Neutron radiography can see very different things than X-rays, because neutrons can pass with ease through lead and steel but are stopped by plastics, water and oils. Neutron sources include radioactive (^{241}Am/ Be and Cf) sources, electrically driven D-T reactions in vacuum tubes and conventional critical nuclear reactors. It might be possible to use a neutron amplifier to increase the neutron flux.

Since the amount of radiation emerging from the opposite side of the material can be detected and measured, variations in this amount (or intensity) of radiation are used to determine thickness or composition of material. Penetrating radiations are those restricted to that part of the electromagnetic spectrum of wavelength less than about 10 nanometres.

SAFETY

Industrial radiographers are in many locations required by governing authorities to use certain types of safety equipment and to work in pairs. The safety equipment usually includes four basic items: a radiation survey meter (such as a Geiger/Mueller counter), an alarming dosimeter or rate meter, a gas-charged dosimeter, and a film badge or thermoluminescent dosimeter (TLD). The easiest way to remember what each of these items does is to compare them to gauges on an automobile.

The survey meter could be compared to the speedometer, as it measures the speed, or rate, at which radiation is being picked up. When properly calibrated, used, and maintained, it allows the radiographer to see the current exposure to radiation at the meter.

It can usually be set for different intensities, and is used to prevent the radiographer from being overexposed to the radioactive source, as well as for verifying the boundary that radiographers are required to maintain around the exposed source during radiographic operations.

The alarming dosimeter could be most closely compared with the tachometer, as it alarms when the radiographer "redlines" or is exposed to too much radiation. When properly calibrated, activated, and worn on the radiographer's person, it will emit an alarm when the meter measures a radiation level in excess of a preset threshold.

This device is intended to prevent the radiographer from inadvertently walking up on an exposed source. The gas-charged dosimeter is like a trip meter in that it measures the total radiation received, but can be reset. It is designed to help the radiographer measure his/her total periodic dose of radiation. When properly calibrated, recharged, and worn on the radiographer's person, it can tell the radiographer at a glance how much radiation to which the device has been exposed since it was last recharged. Radiographers in many states are required to log their radiation exposures and generate an exposure report.

The film badge or TLD is more like a car's odometer. It is actually a specialized piece of radiographic film in a rugged container. It is meant to measure the radiographer's total exposure over time (usually a month) and is used by regulating authorities to monitor the total exposure of certified radiographers in a certain jurisdiction. At the end of the month, the film badge is turned in and is processed. A report of the radiographer's total dose is generated and is kept on file.

When these safety devices are properly calibrated, maintained, and used, it is virtually impossible for a radiographer to be injured by a radioactive overexposure. Sadly, the elimination of just one of these devices can jeopardize the safety of the radiographer and all those who are nearby.

Without the survey meter, the radiation received may be just below the threshold of the rate alarm, and it may be several hours before the radiographer checks the dosimeter, and up to a month or more before the film badge is developed to detect a low intensity overexposure.

Without the rate alarm, one radiographer may inadvertently walk up on the source exposed by the other radiographer. Without the dosimeter, the radiographer may be unaware of an overexposure, or even a radiation burn, which may take weeks to result in noticeable injury.

And without the film badge, the radiographer is deprived of an important tool designed to protect him or her from the effects of a long-term overexposure to occupationally-obtained radiation, and thus may suffer long-term health problems as a result. Industrial radiography appears to have one of the worst safety profiles of the radiation professions, possibly because there are many operators using strong gamma sources (> 2 Ci) in remote sites with little supervision when compared with workers within the nuclear industry or within hospitals. Many of the "lost source" accidents commented on by the International Atomic Energy Agency involve radiography equipment. Lost source accidents have the potential to cause a considerable loss of human life.

One scenario is that a passerby finds the radiography source and not knowing what it is, takes it home. The person shortly afterwards becomes ill and dies as a result of the radiation dose.

The source remains in their home where it continues to irradiate other members of the household. Such an event occurred in March 1984 in Casablanca (Mohammedia) which is part of Morocco; this is related to the more famous Goiβnia accident, where a related chain of events caused members of the public to be exposed to radiation sources.

Chapter 8

Radiation Therapy

Radiation therapy is the medical use of ionizing radiation as part of cancer treatment to control malignant cells (not to be confused with radiology, the use of radiation in medical imaging and diagnosis). Radiotherapy may be used for curative or adjuvant cancer treatment. It is used as palliative treatment (where cure is not possible and the aim is for local disease control or symptomatic relief) or as therapeutic treatment (where the therapy has survival benefit and it can be curative).

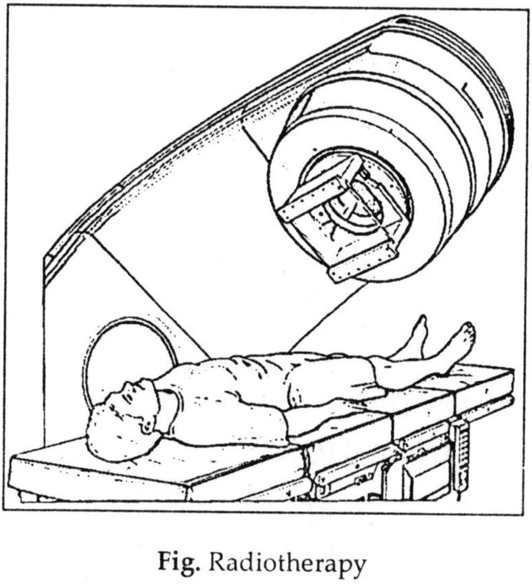

Fig. Radiotherapy

Total body irradiation (TBI) is a radiotherapy technique used to prepare the body to receive a bone marrow transplant.

Radiotherapy has several applications in non-malignant conditions, such as the treatment of trigeminal neuralgia, severe thyroid eye disease, pterygium, pigmented villonodular synovitis, prevention of keloid scar growth, and prevention of heterotopic ossification. The use of radiotherapy in non-malignant conditions is limited partly by worries about the risk of radiation-induced cancers.

Radiotherapy is used for the treatment of malignant tumors (cancer), and may be used as the primary therapy. It is also common to combine radiotherapy with surgery, chemotherapy, hormone therapy or some mixture of the three. Most common cancer types can be treated with radiotherapy in some way. The precise treatment intent (curative, adjuvant, neoadjuvant, therapeutic, or palliative) will depend on the tumour type, location, and stage, as well as the general health of the patient.

Radiation therapy is commonly applied to the cancerous tumour. The radiation fields may also include the draining lymph nodes if they are clinically or radiologically involved with tumour, or if there is thought to be a risk of subclinical malignant spread. It is necessary to include a margin of normal tissue around the tumour to allow for uncertainties in daily set-up and internal tumor motion. These uncertainties can be caused by internal movement (for example, respiration and bladder filling) and movement of external skin marks relative to the tumour position.

To spare normal tissues (such as skin or organs which radiation must pass through in order to treat the tumour), shaped radiation beams are aimed from several angles of exposure to intersect at the tumour, providing a much larger absorbed dose there than in the surrounding, healthy tissue.

MECHANISM OF ACTION

Radiation therapy works by damaging the DNA of cells. The damage is caused by a photon, electron, proton, neutron, or ion beam directly or indirectly ionizing the atoms which make up the DNA chain. Indirect ionization happens as a result of the ionization of water, forming free radicals, notably hydroxyl radicals, which then damage the DNA. In the most common forms of radiation therapy, most of the radiation effect is through free radicals.

Because cells have mechanisms for repairing DNA damage, breaking the DNA on both strands proves to be the most significant technique in modifying cell characteristics. Because cancer cells generally are undifferentiated and stem cell-like, they reproduce more, and have a diminished ability to repair sub-lethal damage compared to most healthy differentiated cells. The DNA damage is inherited through cell division, accumulating damage to the cancer cells, causing them to die or reproduce more slowly. Proton radiotherapy works by sending protons with varying kinetic energy to precisely stop at the tumor.

One of the major limitations of radiotherapy is that the cells of solid tumors become deficient in oxygen. This is because solid tumours usually outgrow their blood supply, causing a low-oxygen state known as hypoxia. The more hypoxic the tumours are the more resistant they are to the effects of radiation because oxygen makes the radiation damage to DNA permanent.

Much research has been devoted to overcoming this problem including the use of high pressure oxygen tanks, blood substitutes that carry increased oxygen, hypoxic cell radiosensitizers such as misonidazole and metronidazole, and hypoxic cytotoxins, such as tirapazamine. There is also interest in the fact that high-LET (linear energy transfer) particles such as carbon or neon ions may have an antitumour effect which is independent of tumour hypoxia.

SIDE EFFECTS

Radiation therapy is in itself painless. Many low-dose palliative treatments (for example, radiotherapy to bony metastases) cause minimal or no side effects. Treatment to higher doses causes varying side effects during treatment (acute side effects), in the months or years following treatment (long-term side effects), or after re-treatment (cumulative side effects). The nature, severity, and longevity of side effects depends on the organs that receive the radiation, the treatment itself (type of radiation, dose, fractionation, concurrent chemotherapy), and the patient.

Most side effects are predictable and expected. Side effects from radiation are usually limited to the area of the patients body that is under treatment. One of the aims of modern radiotherapy is to reduce side effects to a minimum, and to help the patient to understand and to deal with those side effects which are unavoidable.

Acute Side Effects

Damage to the Epithelial Surfaces

Epithelial surfaces like skin, oral, pharyngeal and bowel mucosa, urothelium, etc. may sustain damage from radiation therapy. The rates of onset of damage and recovery from it depend upon the turnover rate of epithelial cells. Typically the skin starts to become pink and sore several weeks into treatment. The reaction may become more severe during the treatment and for up to about one week following the end of radiotherapy, and the skin may break down. Although this moist desquamation is uncomfortable, recovery is usually quick. Skin reactions tend to be worse in areas where there are natural folds in the skin, such as underneath the female breast, behind the ear, and in the groin.

Similarly, the lining of the mouth, throat, esophagus, and bowel may be damaged by radiation. If the head and neck area is treated, temporary soreness and ulceration commonly occur in the mouth and throat. If severe, this can affect swallowing, and the patient may need painkillers and nutritional support. The esophagus can also become sore if it is treated directly, or if, as commonly occurs, it receives a dose of collateral radiation during treatment of lung cancer.

The lower bowel may be treated directly with radiation (treatment of rectal or anal cancer) or be exposed by radiotherapy to other pelvic structures (prostate, bladder, female genital tract). Typical symptoms are soreness, diarrhoea, and nausea.

Swelling (Edema or Oedema)

As part of the general inflammation that occurs, swelling of soft tissues may cause problems during radiotherapy. This is a concern during treatment of brain tumours and brain metastases, especially where there is pre-existing raised intracranial pressure or where the tumour is causing near-total obstruction of a lumen (e.g., trachea or main bronchus). Surgical intervention may be considered prior to treatment with radiation.

If surgery is deemed unnecessary or inappropriate, the patient may receive steroids during radiotherapy to reduce swelling.

Infertility

The gonads (ovaries and testicles) are very sensitive to radiation. They may be unable to produce gametes following direct exposure to most normal treatment doses of radiation. Treatment planning for all body sites is designed to minimize, if not completely exclude dose to the gonads if they are not the primary area of treatment.

MEDIUM AND LONG-TERM SIDE EFFECTS

These depend on the tissue that received the treatment; they may be minimal.

- Fibrosis: Tissues which have been irradiated tend to become less elastic over time due to a diffuse scarring process.
- Hair loss: This may be most pronounced in patients who have received radiotherapy to the brain. Unlike the hair loss seen with chemotherapy, radiation-induced hair loss is more likely to be permanent, but is also more likely to be limited to the area treated by the radiation.
- Dryness: The salivary glands and tear glands have a radiation tolerance of about 30 Gy in 2 Gy fractions, a dose which is exceeded by most radical head and neck cancer treatments. Dry mouth (xerostomia) and dry eyes (xerophthalmia) can become irritating long-

term problems and severely reduce the patient's quality of life. Similarly, sweat glands in treated skin (such as the armpit) tend to stop working, and the naturally moist vaginal mucosa is often dry following pelvic irradiation.

- Fatigue: Fatigue is among the most common symptoms of Radiation therapy, and can range from a few months, a few years, depending on the quantity of the treatment and cancer type. Lack of Energy, reduced activity and overtired feelings are common symptoms
- Cancer: Radiation is a potential cause of cancer, and secondary malignancies are seen in a very small minority of patients, generally many years after they have received a course of radiation treatment. In the vast majority of cases, this risk is greatly outweighed by the reduction in risk conferred by treating the primary cancer.
- Death: Radiation has potentially excess risk of death from heart disease seen after some past breast cancer RT regimens.

Cumulative Side Effects

Cumulative effects from this process should not be confused with long-term effects—when short-term effects have disappeared and long-term effects are subclinical, reirradiation can still be problematic.

DOSE

The amount of radiation used in radiation therapy is measured in gray (Gy), and varies depending on the type and stage of cancer being treated. For curative cases, the typical dose for a solid epithelial tumor ranges from 60 to 80 Gy, while lymphoma tumors are treated with 20 to 40 Gy.

Preventative (adjuvant) doses are typically around 45 - 60 Gy in 1.8 - 2 Gy fractions (for Breast, Head and Neck cancers respectively.) Many other factors are considered by radiation oncologists when selecting a dose, including whether the patient is receiving chemotherapy, whether radiation therapy is being administered before or after surgery, and the degree of success of surgery.

FRACTIONATION

The total dose is fractionated (spread out over time) for several important reasons. Fractionation allows normal cells time to recover, while tumor cells are generally less efficient in repair between fractions. Fractionation also allows tumor cells that were in a relatively radio-resistant phase of the cell cycle during one treatment to cycle into a sensitive phase of the cycle before the next fraction is given.

Similarly, tumor cells that were chronically or acutely hypoxic (and therefore more radioresistant) may reoxygenate between fractions, improving the tumor cell kill. Fractionation regimes are individualised between different radiotherapy centres and even between individual doctors. In the USA, Australia, and Europe, the typical fractionation schedule for adults is 1.8 to 2 Gy per day, five days a week.

In the northern United Kingdom, fractions are more commonly 2.67 to 2.75 Gy per day, which eases the burden on thinly spread resources in the National Health Service. In some cancer types, prolongation of the fraction schedule over too long can allow for the tumor to begin repopulating, and for these tumor types, including head-and-neck and cervical squamous cell cancers, radiation treatment is preferably completed within a certain amount of time. For children, a typical fraction size may be 1.5 to 1.8 Gy per day, as smaller fraction sizes are associated with reduced incidence and severity of late-onset side effects in normal tissues.

In some cases, two fractions per day are used near the end of a course of treatment. This schedule, known as a concomitant boost regimen or hyperfractionation, is used on tumors that regenerate more quickly when they are smaller. In particular, tumors in the head-and-neck demonstrate this behaviour.

One of the best-known alternative fractionation schedules is Continuous Hyperfractionated Accelerated Radiotherapy

(CHART). CHART, used to treat lung cancer, consists of three smaller fractions per day. Although reasonably successful, CHART can be a strain on radiation therapy departments.

Implants can be fractionated over minutes or hours, or they can be permanent seeds which slowly deliver radiation until they become inactive.

TYPES OF RADIATION THERAPY

Historically, the three main divisions of radiotherapy are external beam radiotherapy (EBRT or XBRT) or teletherapy, brachytherapy or sealed source radiotherapy and unsealed source radiotherapy. The differences relate to the position of the radiation source; external is outside the body, while sealed and unsealed source radiotherapy has radioactive material delivered internally.

Brachytherapy sealed sources are usually extracted later, while unsealed sources may be administered by injection or ingestion. Proton therapy is a special case of external beam radiotherapy where the particles are protons.

Introperative radiotherapy is a special type of radiotherapy that is delivered immediately after surgical removal of the cancer. This method has been employed in breast cancer (TARGeted Introperative radioTherapy), brain tumours and rectal cancers.

CONVENTIONAL EXTERNAL BEAM RADIOTHERAPY

Conventional external beam radiotherapy (2DXRT) is delivered via two-dimensional beams using linear accelerator machines. 2DXRT mainly consists of a single beam of radiation delivered to the patient from several directions: often front or back, and both sides.

Conventional refers to the way the treatment is *planned* or *simulated* on a specially calibrated diagnostic x-ray machine known as a simulator because it recreates the linear accelerator actions (or sometimes by eye), and to the usually well-established arrangements of the radiation beams to achieve a desired *plan*. The aim of simulation is to accurately target or localize the volume which is to be treated. This technique is well established and is generally quick and reliable. The worry is that some high-dose treatments may be limited by the radiation toxicity capacity of healthy tissues which lay close to the target tumor volume.

An example of this problem is seen in radiation of the prostate gland, where the sensitivity of the adjacent rectum limited the dose which could be safely prescribed using 2DXRT planning to such an extent that tumor control may not be easily achievable.

Previous to the invention of the CT, physicians and physicists had limited knowledge about the true radiation dosage delivered to both cancerous and healthy tissue. For this reason, 3-dimensional conformal radiotherapy is becoming the standard treatment for a number of tumor sites.

Virtual Simulation, 3-dimensional Conformal Radiotherapy, and Intensity-modulated Radiotherapy

The planning of radiotherapy treatment has been revolutionized by the ability to delineate tumors and adjacent normal structures in three dimensions using specialized CT and/or MRI scanners and planning software.

Virtual simulation, the most basic form of planning, allows more accurate placement of radiation beams than is possible using conventional X-rays, where soft-tissue structures are often difficult to assess and normal tissues difficult to protect.

An enhancement of virtual simulation is 3-Dimensional Conformal Radiotherapy (3DCRT), in which the profile of each radiation beam is shaped to fit the profile of the target from a beam's eye view (BEV) using a multileaf collimator (MLC) and a variable number of beams. When the treatment volume conforms to the shape of the tumour, the relative toxicity of radiation to the surrounding normal tissues is reduced, allowing a higher dose of radiation to be delivered to the tumor than conventional techniques would allow.

Intensity-Modulated Radiation Therapy (IMRT) is an advanced type of high-precision radiation that is the next generation of 3DCRT. IMRT also improves the ability to conform the treatment volume to concave tumor shapes, for example when the tumor is wrapped around a vulnerable structure such as the spinal cord or a major organ or blood vessel.Computer-controlled x-ray accelerators distribute precise radiation doses to malignant tumors or specific areas within the tumor.

The pattern of radiation delivery is determined using highly-tailored computing applications to perform optimization and treatment simulation (Treatment Planning). The radiation dose is consistent with the 3-D shape of the tumor by controlling, or modulating, the radiation beam's intensity.

The radiation dose intensity is elevated near the gross tumor volume while radiation among the neighboring normal tissue is decreased or avoided completely.

The customized radiation dose is intended to maximize tumor dose while simultaneously protecting the surrounding normal tissue. Because sparing healthy tissue as compared with conventional radiation therapy techniques (2DXRT and 3DCRT). This in turn results in better tumor targeting, lessened side effects, and improved treatment outcomes than even 3DCRT.

3DCRT is still used extensively for many body sites but the use of IMRT is growing in more complicated body sites such as CNS, head and neck, prostate, breast and lung. Unfortunately, IMRT is limited by its need for additional time from experienced medical personnel. This is because physicians must manually delineate the tumors one CT image at a time through the entire disease site which can take much longer than 3DCRT preparation.

Then, medical physicists and dosimetrists must be engaged to create a viable treatment plan. Also, the IMRT technology has only been used commercially since the late 1990s even at the most advanced cancer centers, so radiation oncologists who did not learn it as part of their residency programme must find additional sources of education before implementing IMRT. Proof of improved survival benefit from either of these two techniques over conventional radiotherapy (2DXRT) is growing for many tumor sites, but the ability to reduce toxicity is generally accepted. Both techniques enable dose escalation, potentially increasing usefulness. There has been some concern, particularly with 3DCRT, about increased exposure of normal tissue to radiation and the consequent potential for secondary malignancy.

Overconfidence in the accuracy of imaging may increase the chance of missing lesions that are invisible on the planning scans (and therefore not included in the treatment plan) or that move between or during a treatment (for example, due to respiration or inadequate patient immobilization).

New techniques are being developed to better control this uncertainty—for example, real-time imaging combined with real-time adjustment of the therapeutic beams. This new technology is called image-guided radiation therapy (IGRT) or four-dimensional radiotherapy.

RADIOISOTOPE THERAPY (RIT)

Radiotherapy can also be delivered through infusion (into the bloodstream) or ingestion. Examples are the infusion of metaiodobenzylguanidine (MIBG) to treat neuroblastoma, of oral iodine-131 to treat thyroid cancer or thyrotoxicosis, and of hormone-bound lutetium-177 and yttrium-90 to treat neuroendocrine tumors (peptide receptor radionuclide therapy).

Another example is the injection of radioactive glass or resin microspheres into the hepatic artery to radioembolize liver tumors or liver metastases.

In 2002, the United States Food and Drug Administration (FDA) approved Ibritumomab tiuxetan (Zevalin), which is a monoclonal antibody anti-CD20 conjugated to a molecule of Yttrium-90. In 2003, the FDA approved Tositumomab Iodine-131 (Bexxar), which conjugates a molecule of Iodine-131 to the monoclonal antibody anti-CD20.

These medications were the first agents of what is known as radioimmunotherapy, and they were approved for the treatment of refractory non-Hodgkins lymphoma.

Chapter 9

Features of Radioactivity

Radiation is all around us. It comes from the Earth and from outer space. Many forms of radiation are invisible — we can't feel it, see it, taste it, or smell it. Yet, it can be detected and measured when present. We measure ionizing radiation in units called millirems. But what is radiation? Radioactive materials are composed of atoms that are unstable. An unstable atom gives off its excess energy until it becomes stable. The energy emitted is radiation. We can classify radiation as being either natural or man-made.

The Earth is surrounded by radiation. Every day, for example, we are exposed to radon, a radioactive gas from uranium found in soil dispersed in the air; from radioactive potassium in our food and water; from uranium, radium, and thorium in the Earth's crust; and from cosmic rays and the sun.

Fig. Uses of Radiation.

These types of radiation are called natural or background radiation. We are exposed to an average of 300 millirems of natural radiation each year (a millirem is a unit of measure for exposure to radiation). This amounts to natural radiation accounting for nearly 85 per cent of our total annual exposure.

Where does the remaining 15 per cent come from? Man-made sources. Man-made radiation sources that people can be exposed to include tobacco, television, medical x-rays, smoke detectors, lantern mantles, nuclear medicine, and building materials.

Generally, when we think of exposure to radiation, we need to look at radioactive atoms produced in nuclear reactors and described as being unstable. They are unstable because they undergo a disintegrating process called decaying. During this process, unstable atoms becomes stable, throwing off (emitting) radiation in the form of rays and/or particles.

How fast a radioactive atom decays into a stable atom depends on the atom itself. For example, the range in the rate of decay among isotopes goes from fractions of seconds to several billion years (e.g., uranium). Let's take a look at uranium-238 to illustrate the decay chain.

U_{238} →
Thorium$_{230}$ →
Radium$_{226}$ →
Radon$_{218}$ →
Bismuth$_{214}$ →
Lead$_{206}$

As U-238 decays it changes into thorium-230, which changes into radium-226, which changes into radon-218, which changes into bismuth-214, and finally into lead-206 (a stable element).

One peculiar thing about radioactive atoms is that no one knows exactly when the element will decay and give off radiation. There is, however, a pattern relating to how long it takes for an isotope to lose half of its radioactivity. The pattern is called half-life.

If an atom, for example, has a half-life of 10 years, half of its atoms will decay in 10 years. Then in another 10 years half of that amount will decay and so on. While there are several different forms of radiation, we're going to concentrate on just three that result from the decay of radioactive isotopes: alpha, beta, and gamma.

Beta particles are high energy electrons. Both alpha and beta particles are emitted from unstable isotopes. The alpha particle, consisting of two protons and two neutrons, is relatively large compared to beta particles. Gamma rays have no mass.

Because of its size and electrical charge (+2), the alpha particle has a relatively slow speed and low penetrating distance (one to two inches in air). Alpha particles are easily stopped by a thin sheet of paper or the body's outer layer of skin. Since they do not penetrate the outer (dead) layer of skin, they present little or no hazard when they are external to the body.

However, alpha particles are considered internal hazards, because when they come into contact with live tissue they cause a large number of ionizations to occur in small areas, thus causing damage to tissues and cells.

Beta radiation, while faster and lighter than alpha radiation, can travel through about 10 feet of air and penetrate very thin layers of materials such as aluminum foil. However, while clothing will stop most beta particles, they can penetrate the live layers of skin tissue. Therefore, beta radiation is considered to be both an internal and external (to skin only) hazard. Thin layers of metals and plastics can be used to shield individuals from beta radiation.

Gamma radiation, high energy light, is a little different. It is a type of electromagnetic wave, just like radio waves, light waves, and x-rays. Gamma radiation is a very strong type of electromagnetic wave. It is has no weight and travels at the speed of light. This is much faster than alpha and beta radiation.

Because of their penetrating capability, gamma rays are considered both internal and external hazards. Thick walls of cement, lead, or steel are needed to stop it. Alpha, beta, and gamma radiations are also known as ionizing radiation. Ionizing radiation is especially harmful because it can change the chemical makeup of many things, including the delicate chemistry of the human body and other living organisms. For this reason it is a good idea to avoid unnecessary exposure to all ionizing radiation. The problem is simply this: large amounts of radiation — far above the levels encountered in daily life — can produce cancer and genetic defects in living organisms. Radiation causes damage and alters the body's normal cells and normal cell function.

This breakdown in normal cell function may result in an uncontrolled growth of cells, hence the potential for malignant/ cancerous tumors. Whether the source of radiation is natural or man-made, whether it is a small dose of radiation or a large dose, there will be some biological effects.

A diagram can show the biological effect of ionizing radiation. Radiation causes ionizations of atoms that will affect molecules that may affect cells that may effect tissues that may affect organs that may affect the whole body.

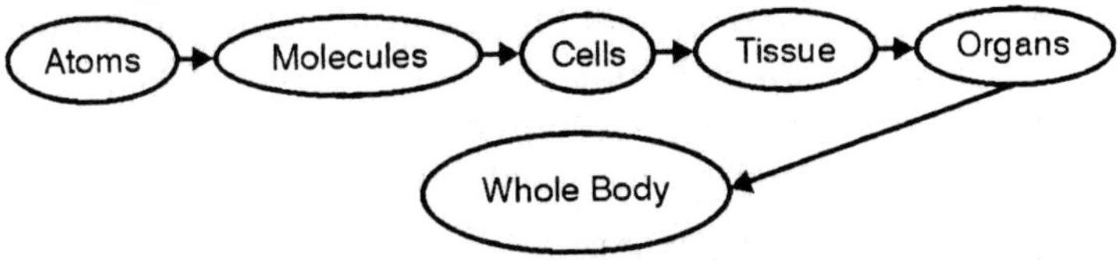

Although we tend to think of biological effects in terms of the effect of radiation on living cells, in actuality, ionizing radiation, by definition, interacts only with atoms by a process called ionization. Thus, all biological damage effects begin when radiation interacts with atoms forming the cells in the human body. As a result, radiation effects on humans proceed from the lowest to the highest level as noted on the board.

Although scientists have only known about radiation since the 1890s, they have developed a wide variety of uses for this natural force. Today, to benefit mankind, radiation is used in science, medicine, and industry, as well as for generating electricity. Radiation has useful applications in such areas as agriculture, medicine, space exploration, architect/engineering, industry/manufacturing, government, geology (including mining), ecology, and education.

Radiation is used by doctors to diagnose illness and helps archaeologists find the age of ancient artifacts. Electricity produced by nuclear fission — splitting the atom — is one of its greatest uses. A reliable source of electricity is needed to give us light, help to groom and feed us, and to keep our homes and businesses running. Some specific examples of how the radiation has been used to-

- Diagnose and treat illnesses.
- Kill bacteria and preserve food without chemicals and refrigeration.
- Process sludge for fertilizer and soil conditioner.
- Locate underground natural resources and tell a dry hole from a gusher.
- Make smoke detectors, nonstick fry pans, and ice cream.
- Grow stronger crops.
- Power satellites and provide future electrical needs for space laboratories with people on board.
- Design, instruments, techniques, and equipment; measure air pollution.
- Prove the age of works of art and assist in determining their authenticity.

RADIATION IN SCIENCE

Radiation is used in science in many ways. Just as doctors can label substances inside people's bodies, scientists can label substances that pass through plants, animals, or our world. This allows us to study such things as the paths that different types of air and water pollution take through the environment.

It has also helped us learn more about a wide variety of things, such as what types of soil different plants need to grow, the size of newly discovered oil fields, and the track of ocean currents.

Scientists use radioactive substances to find the age of ancient objects by a process called carbon dating. For example, in the upper levels of our atmosphere, cosmic rays hit nitrogen atoms and form a naturally radioactive isotope called carbon-14. Carbon is found in all living things, and a small percentage of this carbon is carbon-14.

When a plant or animal dies, it no longer takes in new carbon and the carbon-14 it contains begins the process of radioactive decay. However, new isotopes of carbon-14 continue to be formed in our atmosphere, and after a few years the per cent of radioactivity in an old object is less than it is in a newer one. By measuring this difference, scientists are able to determine how old certain objects are. The measuring process is called carbon dating.

Uranium is an Element that Can be Found in the Crust of the Earth

This element, quite abundant in many areas of the world, is naturally radioactive. Certain isotopes of uranium can be used as fuel in a nuclear power plant. The uranium is formed into ceramic pellets about the size of the end of our finger. These pellets are inserted into long, vertical tubes (fuel rods) within the reactor. The reactor is the heart of the nuclear power plant. Basically, it is a machine that heats water.

A reactor has four main parts: the uranium fuel assemblies, the control rods, the coolant/moderator, and the pressure vessel. The fuel assemblies, control rods, and coolant/moderator make up what is known as the reactor core. The core is surrounded by the pressure vessel.

We also have to understand that uranium cannot just be thrown into a reactor the way we shovel coal into a furnace. The fuel rods, containing the uranium, are carefully bound together into fuel assemblies, each of which contains about 240 rods. The assemblies hold the rods apart so that when they are submerged into the reactor core, water can flow between them.

When the uranium atom splits, it releases energy and two or more neutrons from its nucleus. These neutrons can then hit the nuclei of other uranium atoms causing them to fission. . The sequence of one fission triggering others, and those triggering still more, is called a chain reaction. When the atoms split, they release energy in the form of heat.

The heat is transferred from the reactor core to the steam generator. Here, it's converted to high pressure steam which turns the turbine in the electric generator. This next transparency shows the basic design of the nuclear fuel steam plant and its various components.

In the past few minutes I've introduced several new terms. Let me explain what each means in the energy generating process. First, control rods. What are they? How are they used?

The control rods slide up and down in between the fuel assemblies in the reactor core. They control or regulate the speed of the nuclear reaction by absorbing neutrons. Here's how it works: When the control rods absorb neutrons, fewer neutrons hit the uranium atoms thus slowing down the chain reaction.

On the other hand, when the core temperature goes down, the control rods are slowly lifted out of the core, and fewer neutrons are absorbed. Therefore, more neutrons are available to cause fission. This releases more heat energy.

Just as there are different types of houses and cars, there are different types of nuclear power plants that generate electricity. The two basic types being used are the boiling water reactor (BWR) and the pressurized water reactor (PWR). These power plants are often referred to as light water reactors.

BOILING WATER REACTOR (BWR)

The boiling water reactor operates in essentially the same way as a fossil fuel generating plant. Neither of these types of power plants have a steam generator. Instead, water in the BWR boils inside the pressure vessel and the steam water mixture is produced when very pure water (reactor coolant) moves upward through the core absorbing heat. The water boils and produces steam. When the steam rises to the top of the pressure vessel, water droplets are removed, the steam is sent to the turbine generator to turn the turbine.

PRESSURIZED WATER REACTOR (PWR)

The pressurized water reactor differs from the BWR in that the steam to run the turbine is produced in a steam generator. Water boils at 212°F or 100°C. If a lid is tightly placed over a pot of boiling water (a pressure cooker), the pressure inside the pot will increase because the steam cannot escape. As the pressure increases, so does the temperature of the water in the pot. In the PWR plant, a pressurizer unit keeps the water that is flowing through the reactor vessel under very high pressure to prevent it from boiling. The hot water then flows into the steam generator where it is converted to steam. The steam passes through the turbine which produces electricity. About two-thirds of the reactor power plants in the U.S. are of the PWR type.

MEDICAL USES OF RADIOACTIVITY

Since the discovery of radiation, people have benefited from the use of radiation in medicine, agriculture and industry. Physicians use x-rays in more than half of all medical diagnoses to determine the extent of disease or physical injury. The radioactive isotope cobalt-60 helps to stop the body's immune reaction to transplanted human organs. Also, tests using nuclear materials in hospital laboratories can detect thyroid underactivity in newborn babies. This makes prompt treatment possible, saving many children from mental retardation.

The discovery of x-rays in 1895 was a major turning point in diagnosing diseases because physicians finally had an easy way to "see" inside the body without having to operate. Newer x-ray technologies such as CT (computerized tomography) scans have revolutionized the diagnosis and treatment of diseases affecting almost every part of the body. Other sophisticated techniques have provided physicians with low-risk ways to diagnose heart disease. For example, doctors can now pinpoint cholesterol deposits that are narrowing or blocking coronary arteries, information essential for bypassing or unclogging them.

Every major hospital in the United States has a nuclear medicine department. Radionuclides are used safely and effectively to diagnose and treat a wide variety of diseases more effectively and safely by "seeing" how the disease process alters the normal function of an organ To obtain this information a patient swallows, inhales, or receives an injection of a tiny amount of a radionuclide. Special cameras reveal where the radioactivity accumulates briefly in the body, providing, for example, an image of the heart that shows normal and malfunctioning tissue. Radionuclides are also used in laboratory tests to measure important substances in the body, such as thyroid hormone. Radionuclides are used to effectively treat patients with thyroid diseases, including Graves disease, one of the most common forms of hyperthyroidism, and thyroid cancer.

The use of ionizing radiation has led to major improvements in the diagnosis and treatment of patients with cancer. These innovations have resulted in increased survival rates and improved quality of life. Mammography can detect breast cancer at an early stage when it may be curable. Needle biopsies are more safe, accurate, and informative when guided by x-ray or other imaging techniques.

Radiation is used in monitoring the response of tumors to treatment and in distinguishing malignant tumors from benign ones. Bone and liver scans can detect cancers that have spread to these organs. Half of all people with cancer are treated with radiation, and the number of those who have been cured continues to rise.

There are now tens of thousands of individuals alive and cured from various cancers as a result of radiotherapy. In addition, there are many patients who have entered remission as a result of radiotherapy. Radionuclides are also being used to decrease or eliminate the pain associated with cancer, such as that of the prostate or breast cancer that has spread to the bone.

Radionuclides are a technological backbone for much of the biomedical research being done today. They are used in identifying and learning how genes work. Much of the research on AIDS is dependent upon the use of radionuclides. Scientists are also "arming" monoclonal antibodies, antibodies that are produced in the laboratory and engineered to bind to a specific protein on a patient's tumor cells, with radionuclides.

When such "armed" antibodies are injected into a patient, they bind to the tumor cells, which are then killed by the attached radioactivity, but the nearby normal cells are spared. So far, this approach has produced encouraging success in treating patients with leukemia.

Another clinical and research tool, PET scanning (positron emission tomography), involves injecting radioactive material into a person to "see" the metabolic activity and circulation in a living brain. PET studies have enabled scientists to pinpoint the site of brain tumors or the source of epileptic activity, and to better understand many neurological diseases. For example, researchers were able to learn how dopamine, the chemical messenger (neurotransmitter) that is involved in Parkinson's disease, is used by the brain.

RADIATION IN MEDICINE

X-rays are a type of radiation that can pass through our skin. Our bones are denser than our skin, so when x-rayed, bones and other dense materials cast shadows that can be detected on photographic film. The effect is similar to placing a pencil behind a piece of paper and holding them in front of a light. The shadow of the pencil is revealed because most light has enough energy to pass through the paper, while the denser pencil stops all the light. The difference is that we need film to see the x-rays for us.

Today, doctors and dentists use x-rays to see structures inside our bodies. This allows them to spot broken bones and dental problems. X-ray machines have now been connected to computers in the development of machines called CAT scanners. These instruments provide doctors with Colour TV pictures that show the shape of internal organs.

Approximately 10 million nuclear medicine procedures are performed in the United States annually. Diagnostic x-rays and or radiation therapy were administered to about seven out of every 10 Americans. Medical procedures using radiation have saved thousands of lives through the detection and treatment of conditions ranging from hyperthyroidism to bone cancer.

In such procedures, doctors administer slightly radioactive substances to patients, which are attracted to certain internal organs such as the pancreas, kidney, thyroid, liver, or brain, to diagnose clinical conditions. Moreover, radiation is often used to treat certain types of cancer.

Radioactive iodine, specifically iodine-131, is being used frequently to treat thyroid cancer, a disease which strikes about 11,000 Americans every year

USES OF RADIATION: FOOD IRRADIATION

Food irradiation involves exposing the food, either packaged or in bulk, to carefully controlled amounts of ionizing radiation for a specific time to achieve certain desirable objectives. The process cannot increase the normal radioactivity level of the food, regardless of how long the food is exposed to the radiation, or how much of an energy "dose" is absorbed.

It can prevent the division of living cells, such as bacteria and cells of higher organisms, by changing their molecular structure. It can also slow down ripening or maturation of certain fruits and vegetables by causing biochemical reactions in physiological processes of plant tissues.

FOOD PRESERVATION METHODS

Preservation methods for foods have been known and used for centuries: drying, smoking, curing, and using various simple chemical treatments such as salt, are widespread but they all suffer from the problem that they change the way food looks and tastes. More recently, refrigeration has emerged as the most widely used means of preserving food, or more strictly, for delaying or inhibiting the natural decomposition processes that occur as a result of bacterial action.

It also has the advantage that it has a minimal effect on the taste and consistency of food but it is, relatively speaking, an expensive process. The modern food industry has to make certain choices as to how and when it treats food during the food production cycle.

It can start by reducing the level of micro organisms and pests in food by using chemical treatments and pesticides during growth. For this to be effective the food must then be secured against fresh contamination during transport and storage. An alternative approach is to do very little to the food as grown and harvested, but to treat it nearer to the point of consumption. This is, for example, common with herbs and spices. The food industry will, not surprisingly, tend to choose the way it deals with contamination on the basis of the economics of each case. It will balance the losses in storage and costs of post-treatment against the costs and problems of using more hygienic production and storage methods in the country of origin.

Even where food is produced relatively close to the point of consumption, it may have to be treated because contamination is inherent in the production process. This is why milk has to be pasteurised. Pasteurisation is the most effective way of killing micro organisms with minimal effect on the food itself. Unfortunately pasteurisation can only be used on a very limited range of foods.

NECESSICITY FOOD IRRADIATION

Presently over 40 countries have approved applications to irradiate approximately 40 different foods. These include such items as fruits, vegetables, spices, grains, seafood, meat and poultry. More than half a million tonnes of food is now irradiated throughout the world on a yearly basis. Although this amount represents only a fraction of the food consumed annually, it is constantly growing. This trend is due to three main factors:

- Increasing concerns over food borne diseases.
- High food losses from infestation, contamination and spoilage.
- Growing international trade in food products.

Increasing Concerns Over Foodborne Diseases

Foodborne diseases pose a widespread threat to human health and they are an important cause of reduced economic productivity. Studies by the US Centre for Disease Control in 1999 estimated that foodborne diseases cause approximately 76 million illnesses, 325,000 hospitalizations, and 5,000 deaths in the United States each year. Economic losses associated with such foodborne diseases are high-estimated between US $6.5 billion and $33 billion. The World Health Organization (WHO) has said that food borne diseases are perhaps the most widespread health problem in the contemporary world and an important cause of reduced economic productivity. WHO has also reported that Europe is suffering from an epidemic of salmonellosis.

Pregnant women in the UK have been warned not to eat unpasteurised cheeses which now may have high levels of listeria monocytogenes. Furthermore, even in relatively advanced countries like the UK, the authorities have admitted that the contamination problem is out of control: "The multiplicity of potential routes of contamination makes the elimination of microbiological contamination from poultry being presented for slaughter a virtual impossibility " This need not be the case, however, as has been demonstrated in Sweden.

There it has taken 20 years of ruthless killing of any flock with a salmonella infection to achieve 99 per cent of flocks free of salmonella. Poultry costs more as a result but the Swedish authorities and consumers clearly believe this is worth paying.

High Food Losses from Infestation

The UN Food and Agricultural Authority (FAO) has estimated that about 25 per cent of all worldwide food production is lost after harvesting to insects, bacteria and spoilage. Economic losses due to insects and microbes have been estimated to fall between $5 and $17 billion yearly in the US alone.

Food irradiation can help reduce these losses and can also reduce our dependence on chemical pesticides, some of which are extremely harmful to the environment (e.g. methyl bromide).

Growing International Trade in Food Products

As our economies become more global, food products must meet high standards of quality and quarantine in order to move across borders. Irradiation is an important tool in the fight to prevent the spread of deleterious insects and microorganisms.

AFFECT OF IRRADIATION ON FOOD

The process involves exposing the food, either packaged or in bulk, to carefully controlled amounts of ionizing radiation for a specific time to achieve certain desirable objectives. One objective is the reduction of microorganisms. When microbes present in the food are irradiated, the energy from the radiation breaks the bonds in the DNA molecules, causing defects in the genetic instructions.

Unless this damage can be repaired, the organism will die or will be unable to reproduce. It matters if the food is frozen or fresh, because it takes larger radiation dose to kill microbes in frozen foods. The effectiveness of the process depends also on the organism's sensitivity to irradiation, on the rate at which it can repair damaged DNA, and especially on the amount of DNA in the target organism:

- Parasites and insect pests, which have large amounts of DNA, are rapidly killed by an extremely low dose of irradiation.
- It takes more irradiation to kill bacteria, because they have less DNA.
- Viruses are the smallest pathogens that have nucleic acid, and they are, in general, resistant to irradiation at doses approved for foods.

Another objective is to delay ripening and sprouting. If the food contains living cells, they will be damaged or killed just as microbes are. This is a useful effect: it can be used to prolong the shelf life of fruits and vegetables because it inhibits sprouting and delays ripening.

IRRADIATION COMPARED TO PASTEURIZATION

As in the heat pasteurization of milk, the irradiation process greatly reduces but does not eliminate all bacteria. Irradiated poultry, for example, still requires refrigeration, but would be safe longer than untreated poultry. Strawberries that have been irradiated will last two to three weeks in the refrigerator compared to only a few days for untreated berries. Irradiation complements, but does not replace, the need for proper food handling practices by producers, processors, and consumers. A big advantage of irradiated food, is that it is a cold process: the food is still essentially "raw", because it hasn't undergone any thermal process and it does not significantly increase the temperature or change the physical or sensory characteristics of most foods. An irradiated apple, for example, will still be crisp and juicy. Fresh or frozen meat can be irradiated without cooking it.

APPLICATIONS FOR FOOD IRRADIATION

Because the irradiation process works with both large and small quantities, it has a wide range of potential uses. For example, a single serving of poultry can be irradiated for use on a space flight. Or, a large quantity of potatoes can be treated to reduce sprouting during warehouse storage.

However, irradiation cannot be used with all foods. It causes undesirable flavour changes in dairy products, for example, and it causes tissue softening in some fruits, such as peaches and nectarines (due to breaking of bonds in cellulose and starch). Irradiation is most useful in four areas:

Preservation

Irradiation can be used to destroy or inactivate organisms that cause spoilage and decomposition, thereby extending the shelf life of foods. It is an energy-efficient food preservation method that has several advantages over traditional canning. The resulting products are closer to the fresh state in texture, flavour, and Colour.

Using irradiation to preserve foods requires no additional liquid, nor does it cause the loss of natural juices. Both large and small containers can be used and food can be irradiated after being packaged or frozen.

Sterilization

Foods that are sterilized by irradiation can be stored for years without refrigeration—just like canned (heat sterilized) foods. With irradiation it will be possible to develop new shelf-stable products. Sterilized food is useful in hospitals for patients with severely impaired immune systems, such as some patients with cancer or AIDS. These foods can be used by the military and for space flights.

Control Sprouting, Ripening, and Insect Damage

In this role, irradiation offers an alternative to chemicals for use with potatoes, tropical and citrus fruits, grains, spices, and seasonings. However, since no residue is left in the food, irradiation does not protect against reinfestation like insect sprays and fumigants do.

Control Food Borne Illness

Irradiation can be used to effectively eliminate those pathogens that cause foodborne illness, such as Salmonella.

APPROVED FOR IRRADIATION

A variety of foods have been approved for irradiation, for several different purposes. For meats, separate approval is required both from the FDA and the USDA.

Approval	*Food*	*Purpose*
1963	Wheat flour	Control of mold
1964	White potatoes	Inhibit sprouting
1986	Pork	Kill Trichina parasites
1986	Fruit and vegetables	Insect control Increase shelf life
1986	Herbs and spices	Sterilization
1990 – FDA 1992 -USDA	Poultry	Bacterial pathogen reduction
1997 –FDA 1999 -USDA	Meat	Bacterial pathogen reduction

However, irradiated foods are not widely available yet. Some stores have sold irradiated fruits and vegetables since the early 1990s. Irradiated poultry is available in some grocery stores— mostly small, independent markets— and on menus of a few restaurants.

On the other hand, most spices sold wholesale in this country are irradiated, which eliminates the need for chemical fumigation to control pests. American astronauts have eaten irradiated foods in space since the early 1970s. Patients with weakened immune systems are sometimes fed irradiated foods to reduce the chance of a life-threatening infection. In addition, irradiation is widely used to sterilize a variety of medical and household products, such as joint implants, band-aids, baby pacifiers, cosmetic ingredients, wine and bottle corks, and food packaging materials.

USES OF RADIOACTIVE SUBSTANCES

Carbon-14: Radiometric dating determination of age of carbon-containing artifacts up to about 70,000 years also used as a biological tracer, for example, in studies of photosynthesis. Naturally occurring radioisotope.

Chlorine-36: Measurement of sources of chloride and determining the age of water up to about 2 million years old. Naturally occurring radioisotope.

Lead-210: Date layers of soil and sand deposited up to about 80 years ago. Naturally occurring radioisotope

Tritium: Measure the age of 'young' groundwater up to about 30 years old. Naturally occurring radioisotope. Titriated water is used to study sewage and liquid wastes.

Potassium/ Argon: Radiometric dating 100 000 to several billion years

Rubidium/ Strontium: Radiometric dating:millions of years

Uranium/Lead Radiometric dating:millions to billions of years

Oxygen-18: Biological tracer, for example, in studies of photosynthesis

Sodium-24 Location of leaks in water pipes, studies of body electrolytes. Isotope prepared in a nuclear reactor

Magnesium-27: Location of leaks in water pipes

Potassium-42: Determination of exchanged potassium in

blood flow. Isotope prepared in a nuclear reactor

Chromium-51: Labelling of red blood cells & quantifying gastro-intestinal protein loss. Isotope prepared in a nuclear reactor

Iron-59: In blood studies, when incorporated into steel it is used to determine the amount of friction in machinery

Cobalt-60: Cancer treatment as tumour cells tend to be more susceptible to radiation than other cells

Gallium-67: Tumour-seeking agent. Isotope prepared in a cyclotron

Krypton-81: Lung ventilation studies. Isotope prepared in a cyclotron

Technetium-99: Medical tracer used to locate brain tumours and problems with the lungs, thyroid, liver, spleen, kidney, gall bladder, skeleton, blood pool, bone marrow, salivary & lacrimal glands & heart blood pool & to detect infection. Isotope prepared in a nuclear reactor

Iodine-131: Medical tracer to study & treat the thyroid gland & used in the diagnosis of adrenal medullary & for imaging suspected neural crest and other endocrine tumours. Isotope prepared in a nuclear reactor

Iodine-123: used in imaging to monitor thyroid function & detect adrenal dysfunction. Isotope prepared in a cyclotron

Ytterbium-169: Brain scans. Isotope prepared in a nuclear reactor

Uranium-235: Enriched as a fuel for most nuclear reactors

Plutonium-239: fuel for most "fast-breeder" nuclear reactors

Americium-241: Domestic smoke alarms & neutron gauging. Americium-241 is a decay product of plutonium-241 formed in nuclear reactors.

Copper-64: studying genetic disease affecting copper metabolism. Isotope prepared in a nuclear reactor

Iridium-192: Supplied as a wire for use as an internal radiotherapy device

Molybdenum-99: Used as the 'parent' in a generator to produce technetium-99m, the most widely used radioisotope in nuclear medicine. Isotope prepared in a nuclear reactor

Phosphorus-32: Treatment of excess red blood cells. Isotope prepared in a nuclear reactor

Samarium-153: Used in the treatment of pain associated with bony metastases of primary tumours. Isotope prepared in a nuclear reactor

Yttrium-90: Liver cancer therapy. Isotope prepared in a nuclear reactor

Thallium-201: Locating damaged heart muscle. Isotope prepared in a cyclotron

Caesium-137: Radiotracing to identify sources of soil erosion & depositing, also used in thickness gauging

Gold-198: Trace factory waste causing ocean pollution and to trace sand movement in river beds and on ocean floors

Gadolinium: Used in X-ray fluorescence

Zinc-65 &
Manganese-54 Used to predict the behaviour of heavy metals in effluents from mining waste water

Chapter 10

X-ray

X-radiation (composed of X-rays) is a form of electromagnetic radiation. X-rays have a wavelength in the range of 10 to 0.001 nanometers, corresponding to frequencies in the range 30 petahertz to 30 exahertz (30×10^{15} Hz to 30×10^{18} Hz) and energies in the range 120 eV to 120 keV. They are shorter in wavelength than UV rays. In many languages, X-radiation is called Röntgen radiation after one of its first investigators, Wilhelm Conrad Röntgen.

X-rays are primarily used for diagnostic radiography and crystallography. As a result, the term *X-ray* is metonymically used to refer to a radiographic image produced using this method, in addition to the method itself. X-rays are a form of ionizing radiation and as such can be dangerous.

X-rays span 3 decades in wavelength, frequency and energy. From about 0.12 to 12 keV they are classified as soft X-rays, and from about 12 to 120 keV as hard X-rays, due to their penetrating abilities.

The distinction between X-rays and gamma rays has changed in recent decades. Originally, the electromagnetic radiation emitted by X-ray tubes had a longer wavelength than the radiation emitted by radioactive nuclei (gamma rays). So older literature distinguished between X- and gamma radiation on the basis of wavelength, with radiation shorter than some arbitrary wavelength, such as 10^{-11} m, defined as gamma rays.

However, as shorter wavelength continuous spectrum "X-ray" sources such as linear accelerators and longer wavelength "gamma ray" emitters were discovered, the wavelength bands largely overlapped. The two types of radiation are now usually defined by their origin: X-rays are emitted by electrons outside the nucleus, while gamma rays are emitted by the nucleus.

UNITS OF MEASURE AND EXPOSURE

The measure of X-rays ionizing ability is called the exposure:

- The coulomb per kilogram (C/kg) is the SI unit of ionizing radiation exposure, and measures the amount of radiation required to create 1 coulomb of charge of each polarity in 1 kilogram of matter.
- The röntgen (R) is an obsolete older traditional unit of exposure, which represented the amount of radiation required to create 1 esu of charge of each polarity in 1 cubic centimeter of dry air. 1 roentgen = 2.58×10^{-4} C/kg

However, the effect of ionizing radiation on matter (especially living tissue) is more closely related to the amount of energy deposited rather than the charge. This is called the absorbed dose:

- The gray (Gy) which has units of (J/C), is the SI unit of absorbed dose which is the amount of radiation required to deposit 1 joule of energy in 1 kilogram of any kind of matter.
- The rad is the (obsolete) corresponding traditional unit, equal to 0.01 J deposited per kg. 100 rad = 1 Gy.

The equivalent dose is the measure of the biological effect of radiation on human tissue. For X-rays it is equal to the absorbed dose.

- The sievert (Sv) is the SI unit of equivalent dose, which for X-rays is equal to the gray (Gy).
- The rem is the traditional unit of equivalent dose. For X-rays it is equal to the rad or 0.01 J of energy deposited per kg. 1 sievert = 100 rem.

Medical X-rays are a major source of *manmade* radiation exposure, accounting for 58 per cent in the USA in 1987, but since most radiation exposure is natural (82 per cent) it only accounts for 10 per cent of *total* USA radiation exposure.

Reported dosage due to dental X-rays seems to vary significantly. Depending on the source, a typical dental X-ray of a human results in an exposure of perhaps, 3, 40, 300, or as many as 900 mrems (30 to 9,000 ìSv).

MEDICAL PHYSICS

X-ray K-series spectral line wavelengths (nm) for some common target materials.

X-ray K-series spectral line wavelengths (nm) for some common target materials.				
Target	$K\beta_1$	$K\beta_2$	$K\alpha_1$	$K\alpha_2$
Fe	0.17566	0.17442	0.193604	0.193998
Co				
Ni	0.15001	0.14886	0.165791	0.166175
Cu	0.139222	0.138109	0.154056	0.154439
Zr	0.070173	0.068993	0.078593	0.079015
Mo	0.063229	0.062099	0.070930	0.071359
W				
Re				

X-rays are generated by an X-ray tube, a vacuum tube that uses a high voltage to accelerate electrons released by a hot cathode to a high velocity. The high velocity electrons collide with a metal target, the anode, creating the X-rays.In medical X-ray tubes the target is usually tungsten or a more crack-resistant alloy of rhenium (5 per cent) and tungsten (95 per cent), but sometimes molybdenum for more specialized applications, such as when soft X-rays are needed as in mammography. In crystallography, a copper target is most common, with cobalt often being used when fluorescence from iron content in the sample might otherwise present a problem. The maximum energy of the produced X-ray photon in keV is limited by the energy of the incident electron, which is equal to the voltage on the tube, so an 80 kV tube can't create higher than 80 keV X-rays. When the electrons hit the target, X-rays are created by two different atomic processes:

- *X-ray fluorescence:* If the electron has enough energy it can knock an orbital electron out of the inner shell of a metal atom, and as a result electrons from higher energy levels then fill up the vacancy and X-ray photons are emitted.

This process produces a discrete spectrum of X-ray frequencies, called spectral lines. The spectral lines generated depends on the target (anode) element used and thus are called characteristic lines. Usually these are transitions from upper shells into K shell (called K lines), into L shell (called L lines) and so on.

- *Bremsstrahlung*: This is radiation given off by the electrons as they are scattered by the strong electric field near the high-Z (proton number) nuclei. These X-rays have a continuous spectrum. The intensity of the X-rays increases linearly with decreasing frequency, from zero at the energy of the incident electrons, the voltage on the X-ray tube.

So the resulting output of a tube consists of a continuous bremsstrahlung spectrum falling off to zero at the tube voltage, plus several spikes at the characteristic lines. The voltages used in diagnostic X-ray tubes, and thus the highest energies of the X-rays, range from roughly 20 to 150 kV.

In medical diagnostic applications, the low energy (soft) X-rays are unwanted, since they are totally absorbed by the body, increasing the dose. So a thin metal (often aluminum) sheet is placed over the window of the X-ray tube, filtering out the low energy end of the spectrum. This is called *hardening* the beam.

Both X-ray production processes are extremely inefficient (~1 per cent) and thus to produce a usable flux of X-rays plenty of energy has to be wasted into heat, which has to be removed from the X-ray tube.

Radiographs obtained using X-rays can be used to identify a wide spectrum of pathologies. Due to their short wavelength, in medical applications, X-rays act more like a particle than a wave. This is in contrast to their application in crystallography, where their wave-like nature is most important.

To take an X-ray of the bones, short X-ray pulses are shot

through a body with radiographic film behind. The bones absorb the most photons by the photoelectric process, because they are more electron-dense. The X-rays that do not get absorbed turn the photographic film from white to black, leaving a white shadow of bones on the film.

To generate an image of the cardiovascular system, including the arteries and veins (angiography) an initial image is taken of the anatomical region of interest. A second image is then taken of the same region after iodinated contrast material has been injected into the blood vessels within this area. These two images are then digitally subtracted, leaving an image of only the iodinated contrast outlining the blood vessels. The radiologist or surgeon then compares the image obtained to normal anatomical images to determine if there is any damage or blockage of the vessel.

A specialized source of X-rays which is becoming widely used in research is synchrotron radiation, which is generated by particle accelerators. Its unique features are brightness many orders of magnitude greater than X-ray tubes, wide spectrum, high collimation, and linear polarization.

DETECTORS

PHOTOGRAPHIC PLATE

The detection of X-rays is based on various methods. The most commonly known methods are a photographic plate, X-ray film in a cassette, and rare earth screens. Regardless of what is "catching" the image, they are all categorized as "Image Receptors" (IR).

Before computers and before digital imaging, a photographic plate was used to produce radiographic images. The images were produced right on the glass plates. Film replaced these plates and was used in hospitals to produce images. Now computed & digital radiography has started to replace film in medicine, though film technology remains in use in industrial radiography processes (e.g. to inspect welded seams). Photographic plates are a thing of history, and their replacement (intensifying screens) is now becoming part of that same history. Silver (necessary to the radiographic & photographic industry) is a non-renewable resource, that has now been replaced by digital (DR) and computed (CR) technology.

Where film required wet processing facilities, these new technologies do not. Archiving of these new technologies also saves space. Since photographic plates are sensitive to X-rays, they provide a means of recording the image, but require a lot of exposure (to the patient), so intensifying screens were devised.

They allow a lower dose to the patient, because the screens take the X-ray information and intensify it so that it can be recorded on film positioned next to the intensifying screen. The part of the patient to be X-rayed is placed between the X-ray source and the image receptor to produce a shadow of the internal structure of that particular part of the body.

X-rays are partially blocked ("attenuated") by dense tissues such as bone, and pass more easily through soft tissues. Areas where the X-rays strike darken when developed, causing bones to appear lighter than the surrounding soft tissue. Contrast compounds containing barium or iodine, which are radiopaque, can be ingested in the gastrointestinal tract (barium) or injected in the artery or veins to highlight these vessels.

The contrast compounds have high atomic numbered elements in them that (like bone) essentially block the X-rays and hence the once hollow organ or vessel can be more readily seen. In the pursuit of a non-toxic contrast material, many types of high atomic number elements were evaluated. For example, the first time the forefathers used contrast it was chalk, and was used on a cadaver's vessels.

Unfortunately, some elements chosen proved to be harmful – for example, thorium was once used as a contrast medium (Thorotrast) – which turned out to be toxic in some cases (causing injury and occasionally death from the effects of thorium poisoning). Modern contrast material has improved, and while there is no way to determine who may have sensitivity to the contrast, the incidence of "allergic-type reactions" are low. (The risk is comparable to that associated with penicillin.)

PHOTOSTIMULABLE PHOSPHORS (PSPS)

An increasingly common method of is the use of photostimulated luminescence (PSL), pioneered by Fuji in the 1980s. In modern hospitals a photostimulable phosphor plate (PSP plate) is used in place of the photographic plate. After the plate is X-rayed, excited electrons in the phosphor material remain "trapped" in "colour centres" in the crystal lattice until stimulated by a laser beam passed over the plate surface.

The light given off during laser stimulation is collected by a photomultiplier tube and the resulting signal is converted into a digital image by computer technology, which gives this process its common name, computed radiography (also referred to as digital radiography). The PSP plate can be reused, and existing X-ray equipment requires no modification to use them.

GEIGER COUNTER

Initially, most common detection methods were based on the ionization of gases, as in the Geiger-Müller counter: a sealed volume, usually a cylinder, with a mica, polymer or thin metal window contains a gas, and a wire, and a high voltage is applied between the cylinder (cathode) and the wire (anode). When an X-ray photon enters the cylinder, it ionizes the gas and forms ions and electrons. Electrons accelerate toward the anode, in the process causing further ionization along their trajectory.

This process, known as a Townsend avalanche, is detected as a sudden current, called a "count" or "event". Ultimately, the electrons form a virtual cathode around the anode wire, drastically reducing the electric field in the outer portions of the tube. This halts the collisional ionizations and limits further growth of avalanches. As a result, all "counts" on a Geiger counter are the same size and it can give no indication as to the particle energy of the radiation, unlike the proportional counter. The intensity of the radiation is measurable by the

Geiger counter as the counting-rate of the system. In order to gain energy spectrum information, a diffracting crystal may be used to first separate the different photons. The method is called wavelength dispersive X-ray spectroscopy (WDX or WDS). Position-sensitive detectors are often used in conjunction with dispersive elements. Other detection equipment that is inherently energy-resolving may be used, such as the aforementioned proportional counters.

In either case, use of suitable pulse-processing (MCA) equipment allows digital spectra to be created for later analysis. For many applications, counters are not sealed but are constantly fed with purified gas, thus reducing problems of contamination or gas aging. These are called "flow counters".

SCINTILLATORS

Some materials such as sodium iodide (NaI) can "convert" an X-ray photon to a visible photon; an electronic detector can be built by adding a photomultiplier. These detectors are called "scintillators", filmscreens or "scintillation counters". The main advantage of using these is that an adequate image can be obtained while subjecting the patient to a much lower dose of X-rays.

IMAGE INTENSIFICATION

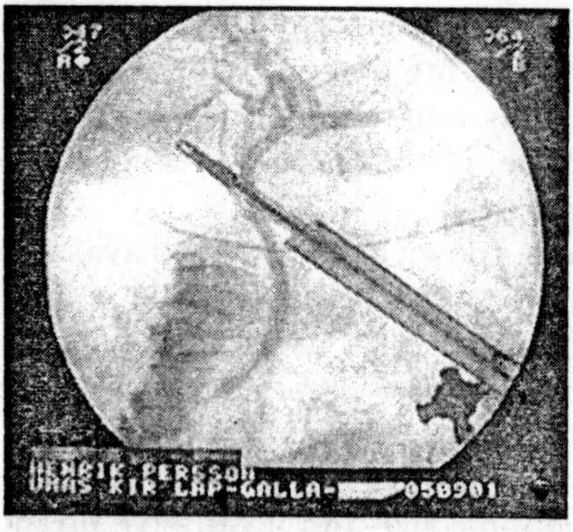

Fig. X-ray During Cholecystectomy

X-rays are also used in "real-time" procedures such as

angiography or contrast studies of the hollow organs (e.g. barium enema of the small or large intestine) using fluoroscopy acquired using an X-ray image intensifier. Angioplasty, medical interventions of the arterial system, rely heavily on X-ray-sensitive contrast to identify potentially treatable lesions.

Direct Semiconductor Detectors

Since the 1970s, new semiconductor detectors have been developed (silicon or germanium doped with lithium, Si(Li) or Ge(Li)). X-ray photons are converted to electron-hole pairs in the semiconductor and are collected to detect the X-rays.

When the temperature is low enough (the detector is cooled by Peltier effect or even cooler liquid nitrogen), it is possible to directly determine the X-ray energy spectrum; this method is called energy dispersive X-ray spectroscopy (EDX or EDS); it is often used in small X-ray fluorescence spectrometers.

These detectors are sometimes called "solid state detectors". Cadmium telluride (CdTe) and its alloy with zinc, cadmium zinc telluride detectors have an increased sensitivity, which allows lower doses of X-rays to be used. Practical application in medical imaging didn't start taking place until the 1990s. Currently amorphous selenium is used in commercial large area flat panel X-ray detectors for mammography and chest radiography. Current research and development is focused around pixel detectors, such as CERN's energy resolving Medipix detector.

Note: A standard semiconductor diode, such as a 1N4007, will produce a small amount of current when placed in an X-ray beam. A test device once used by Medical Imaging Service personnel was a small project box that contained several diodes of this type in series, which could be connected to an oscilloscope as a quick diagnostic.

Silicon drift detectors (SDDs), produced by conventional semiconductor fabrication, now provide a cost-effective and high resolving power radiation measurement. Unlike conventional X-ray detectors, such as Si(Li)s, they do not need to be cooled with liquid nitrogen.

Scintillator Plus Semiconductor Detectors

With the advent of large semiconductor array detectors it has become possible to design detector systems using a scintillator screen to convert from X-rays to visible light which is then converted to electrical signals in an array detector. Indirect Flat Panel Detectors (FPDs) are in widespread use today in medical, dental, veterinary and industrial applications.

The array technology is a variant on the amorphous silicon TFT arrays used in many flat panel displays, like the ones in computer laptops. The array consists of a sheet of glass covered with a thin layer of silicon that is in an amorphous or disordered state. At a microscopic scale, the silicon has been imprinted with millions of transistors arranged in a highly ordered array, like the grid on a sheet of graph paper. Each of these thin film transistors (TFTs) are attached to a light-absorbing photodiode making up an individual pixel (picture element). Photons striking the photodiode are converted into two carriers of electrical charge, called electron-hole pairs. Since the number of charge carriers produced will vary with the intensity of incoming light photons, an electrical pattern is created that can be swiftly converted to a voltage and then a digital signal, which is interpreted by a computer to produce a digital image.

Although silicon has outstanding electronic properties, it is not a particularly good absorber of X-ray photons. For this reason, X-rays first impinge upon scintillators made from e.g. gadolinium oxysulfide or caesium iodide. The scintillator absorbs the X-rays and converts them into visible light photons that then pass onto the photodiode array.

VISIBILITY TO THE HUMAN EYE

While generally considered invisible to the human eye, in special circumstances X-rays can be visible. Brandes, in an experiment a short time after Röntgen's landmark 1895 paper, reported after dark adaptation and placing his eye close to an X-ray tube, seeing a faint "blue-gray" glow which seemed to originate within the eye itself.

Upon hearing this, Röntgen reviewed his record books and found he too had seen the effect. When placing an X-ray tube on the opposite side of a wooden door Röntgen had noted the same blue glow, seeming to emanate from the eye itself, but thought his observations to be spurious because he only saw the effect when he used one type of tube.

Later he realized that the tube which had created the effect was the only one powerful enough to make the glow plainly visible and the experiment was thereafter readily repeatable. The knowledge that X-rays are actually faintly visible to the dark-adapted naked eye has largely been forgotten today; this is probably due to the desire not to repeat what would now be seen as a recklessly dangerous and potentially harmful experiment with ionizing radiation.

It is not known what exact mechanism in the eye produces the visibility: it could be due to conventional detection (excitation of rhodopsin molecules in the retina), direct excitation of retinal nerve cells, or secondary detection via, for instance, X-ray induction of phosphorescence in the eyeball with conventional retinal detection of the secondarily produced visible light.

Though X-rays are invisible it is possible to see the ionization of the air molecules if the intensity of the X-ray beam is high enough. The beamline from the wiggler at the ID11 at ESRF is one example of such high intensity.

MEDICAL USES

Since Röntgen's discovery that X-rays can identify bony structures, X-rays have been developed for their use in medical imaging. Radiology is a specialized field of medicine. Radiologists employ radiography and other techniques for diagnostic imaging.

This is probably the most common use of X-ray technology. X-rays are especially useful in the detection of pathology of the skeletal system, but are also useful for detecting some disease processes in soft tissue.

Some notable examples are the very common chest X-ray, which can be used to identify lung diseases such as pneumonia, lung cancer or pulmonary edema, and the abdominal X-ray, which can detect ileus (blockage of the intestine), free air (from visceral perforations) and free fluid (in ascites).

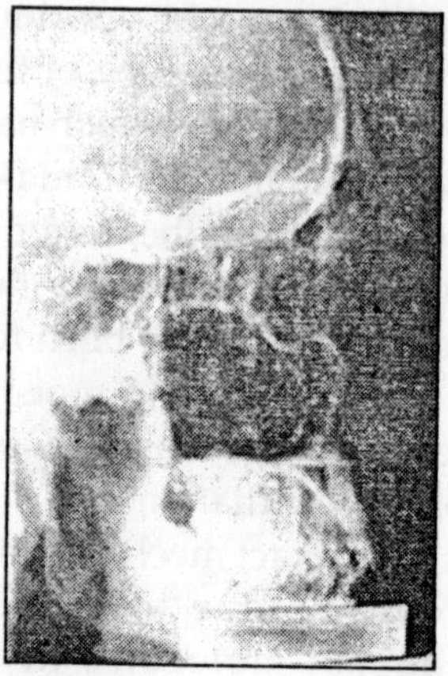

Fig. X-Ray Image of the Paranasal Sinuses, Lateral Projection

In some cases, the use of X-rays is debatable, such as gallstones (which are rarely radiopaque) or kidney stones (which are often visible, but not always). Also, traditional plain X-rays pose very little use in the imaging of soft tissues such as the brain or muscle.

Imaging alternatives for soft tissues are computed axial tomography (CAT or CT scanning), magnetic resonance imaging (MRI) or ultrasound.

Radiotherapy, a curative medical intervention, now used almost exclusively for cancer, employs higher energies of radiation.

SHIELDING AGAINST X-RAYS

Lead is the most common shield agains X-Rays because of its high density (11340 kg/m^3), its easiness of installation and low price. The following table shows the recommended thickness of lead shielding in function of X-Ray energy, from the Recommendations by the Second International Congress of Radiology.

X-Rays generated by peak voltages not exceeding	Minimum thickness of Lead
75 kV	1.0 mm
100 kV	1.5 mm
125 kV	2.0 mm
150 kV	2.5 mm
175 kV	3.0 mm
200 kV	4.0 mm
225 kV	5.0 mm
300 kV	9.0 mm
400 kV	15.0 mm
500 kV	22.0 mm
600 kV	34.0 mm
900 kV	51.0 mm

Other uses

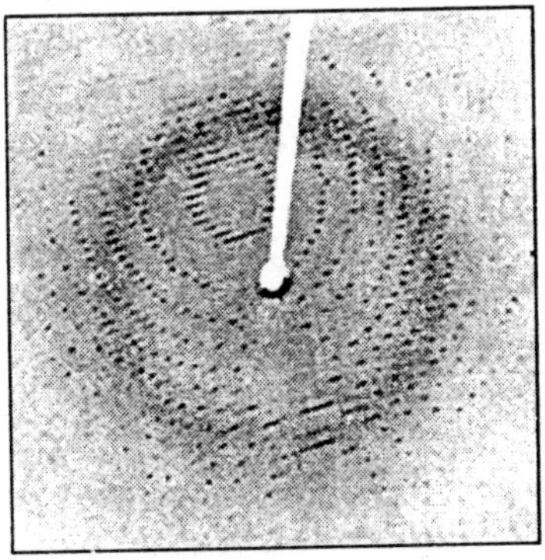

Fig. Scattered X-rays

Each dot, called a reflection, in this diffraction pattern forms from the constructive interference of scattered X-rays passing through a crystal. The data can be used to determine the crystalline structure.

Other Notable Uses of X-rays Include

- X-ray crystallography in which the pattern produced

by the diffraction of X-rays through the closely spaced lattice of atoms in a crystal is recorded and then analyzed to reveal the nature of that lattice. A related technique, fibre diffraction, was used by Rosalind Franklin to discover the double helical structure of DNA.

- X-ray astronomy, which is an observational branch of astronomy, which deals with the study of X-ray emission from celestial objects.
- X-ray microscopic analysis, which uses electromagnetic radiation in the soft X-ray band to produce images of very small objects.
- X-ray fluorescence, a technique in which X-rays are generated within a specimen and detected. The outgoing energy of the X-ray can be used to identify the composition of the sample.
- Industrial radiography uses X-rays for inspection of industrial parts, particularly welds.
- Paintings are often X-rayed to reveal the underdrawing and pentimenti or alterations in the course of painting, or by later restorers. Many pigments such as lead white show well in X-ray photographs.
- Airport security luggage scanners use X-rays for inspecting the interior of luggage for security threats before loading on aircraft.

Wilhelm Röntgen

On November 8, 1895, German physics professor Wilhelm Conrad Röntgen, stumbled on X-rays while experimenting with Lenard and Crookes tubes and began studying them. He wrote an initial report "*On a new kind of ray: A preliminary communication*" and on December 28, 1895 submitted it to the Würzburg's Physical-Medical Society journal.

This was the first paper written on X-rays. Röntgen referred to the radiation as "X", to indicate that it was an unknown type of radiation. The name stuck, although (over Röntgen's great objections), many of his colleagues suggested calling them Röntgen rays. They are still referred to as such in many languages, including German. Röntgen received the first Nobel Prize in Physics for his discovery.

There are conflicting accounts of his discovery because Röntgen had his lab notes burned after his death, but this is a likely reconstruction by his biographers: Röntgen was investigating cathode rays with a fluorescent screen painted with barium platinocyanide and a Crookes tube which he had wrapped in black cardboard so the visible light from the tube wouldn't interfere. He noticed a faint green glow from the screen, about 1 meter away.

The invisible rays coming from the tube to make the screen glow were passing through the cardboard. He found they could also pass through books and papers on his desk. Röntgen threw himself into investigating these unknown rays systematically.

Röntgen discovered its medical use when he saw a picture of his wife's hand on a photographic plate formed due to X-rays. His wife's hand's photograph was the first ever photograph of a human body part using X-rays.

Johann Hittorf

German physicist Johann Hittorf (1824 - 1914), a coinventor and early researcher of the Crookes tube, found when he placed unexposed photographic plates near the tube, that some of them were flawed by shadows, though he did not investigate this effect.

Ivan Pulyui

In 1877 Ukranian-born Pulyui, a lecturer in experimental physics at the University of Vienna, constructed various designs of vacuum discharge tube to investigate their properties.

He continued his investigations when appointed professor at the Prague Polytechnic and in 1886 he found that that sealed photographic plates became dark when exposed to the emanations from the tubes. Early in 1896, just a few weeks after Röntgen published his first X-ray photograph, Pulyui published high-quality x-ray images in journals in Paris and London. Although Pulyui had studied with Röntgen at the University of Strasbourg in the years 1873-75, his biographer Gaida (1997) asserts that his subsequent research was conducted independently. The first medical X-ray made in the United States was obtained using a discharge tube of Pulyui's design.

In January 1896, on reading of Röntgen's discovery, Frank Austin of Dartmouth College tested all of the discharge tubes in the physics laboratory and found that only the Pulyui tube produced X-rays. This was a result of Pulyui's inclusion of an oblique "target" of mica, used for holding samples of fluorescent material, within the tube.

On 3 February 1896 Gilman Frost, professor of medicine at the college, and his brother Edwin Frost, professor of physics, exposed the wrist of Eddie McCarthy, whom Edwin had treated some weeks earlier for a fracture, to the X-rays and collected the resulting image of the broken bone on gelatin photographic plates obtained from Howard Langill, a local photographer also interested in Röntgen's work.

Nikola Tesla

In April 1887, Nikola Tesla began to investigate X-rays using high voltages and tubes of his own design, as well as Crookes tubes. From his technical publications, it is indicated that he invented and developed a special single-electrode X-ray tube, which differed from other X-ray tubes in having no target electrode.

The principle behind Tesla's device is called the Bremsstrahlung process, in which a high-energy secondary X-ray emission is produced when charged particles (such as electrons) pass through matter. By 1892, Tesla performed several such experiments, but he did not categorize the emissions as what were later called X-rays. Tesla generalized the phenomenon as radiant energy of "invisible" kinds. Tesla stated the facts of his methods concerning various experiments in his 1897 X-ray lecture before the New York Academy of Sciences. Also in this lecture, Tesla stated the method of construction and safe operation of X-ray equipment. His X-ray experimentation by vacuum high field emissions also led him to alert the scientific community to the biological hazards associated with X-ray exposure.

Philipp Lenard

Philipp Lenard, a student of Heinrich Hertz, wanted to see whether cathode rays could pass out of the Crookes tube into the air. He built a Crookes tube (later called a "Lenard tube") with a "window" in the end made of thin aluminum, facing the cathode so the cathode rays would strike it.

He found that something came through, that would expose photographic plates and cause fluorescence. He measured the penetrating power of these rays through various materials. It has been suggested that at least some of these "Lenard rays" were actually X-rays.

Hermann von Helmholtz formulated mathematical equations for X-rays. He postulated a dispersion theory before Röntgen made his discovery and announcement. It was formed on the basis of the electromagnetic theory of light. However, he did not work with actual X-rays.

Thomas Edison

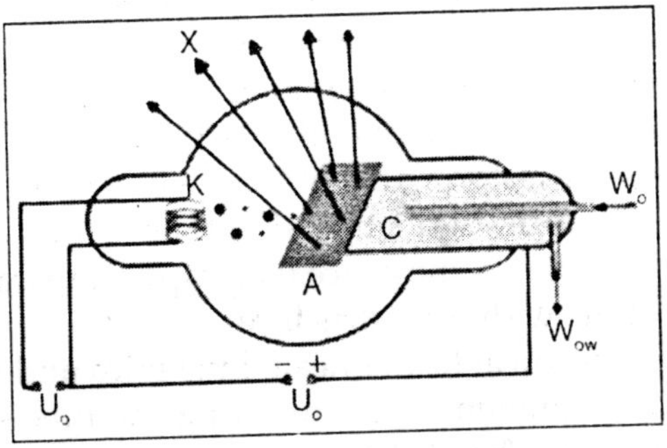

Fig. Diagram of a Water Cooled X-ray Tube.

In 1895, Thomas Edison investigated materials' ability to fluoresce when exposed to X-rays, and found that calcium tungstate was the most effective substance. Around March 1896, the fluoroscope he developed became the standard for medical X-ray examinations. Nevertheless, Edison dropped X-ray research around 1903 after the death of Clarence Madison Dally, one of his glassblowers.

Dally had a habit of testing X-ray tubes on his hands, and acquired a cancer in them so tenacious that both arms were amputated in a futile attempt to save his life. "At the 1901 Pan-American Exposition in Buffalo, New York, an assassin shot President William McKinley twice at close range with a.32 caliber revolver."

The first bullet was removed but the second remained lodged somewhere in his stomach. McKinley survived for some time and requested that Thomas Edison "rush an X-ray machine to Buffalo to find the stray bullet.

The 20th Century and Beyond

The many applications of X-rays immediately generated enormous interest. Workshops began making specialized versions of Crookes tubes for generating X-rays, and these first generation cold cathode or Crookes X-ray tubes were used until about 1920. Crookes tubes were unreliable.

They had to contain a small quantity of gas (invariably air) as a current will not flow in such a tube if they are fully evacuated. However as time passed the X-rays caused the glass to absorb the gas, causing the tube to generate "harder" X-rays until it soon stopped operating.

Larger and more frequently used tubes were provided with devices for restoring the air, known as "softeners". This often took the form of a small side tube which contained a small piece of mica – a substance that traps comparatively large quantities of air within its structure.

A small electrical heater heated the mica and caused it to release a small amount of air restoring the tube's efficiency. However the mica itself had a limited life and the restore process was consequently difficult to control.

In 1904, John Ambrose Fleming invented the thermionic diode valve (vacuum tube). This used a hot cathode which permitted current to flow in a vacuum. This idea was quickly applied X-ray tubes, and heated cathode X-ray tubes, called

Coolidge tubes, replaced the troublesome cold cathode tubes by about 1920.

Two years later, physicist Charles Barkla discovered that X-rays could be scattered by gases, and that each element had a characteristic X-ray. He won the 1917 Nobel Prize in Physics for this discovery. Max von Laue, Paul Knipping and Walter Friedrich observed for the first time the diffraction of X-rays by crystals in 1912.

This discovery, along with the early works of Paul Peter Ewald, William Henry Bragg and William Lawrence Bragg gave birth to the field of X-ray crystallography. The Coolidge tube was invented the following year by William D. Coolidge which permitted continuous production of X-rays; this type of tube is still in use today.

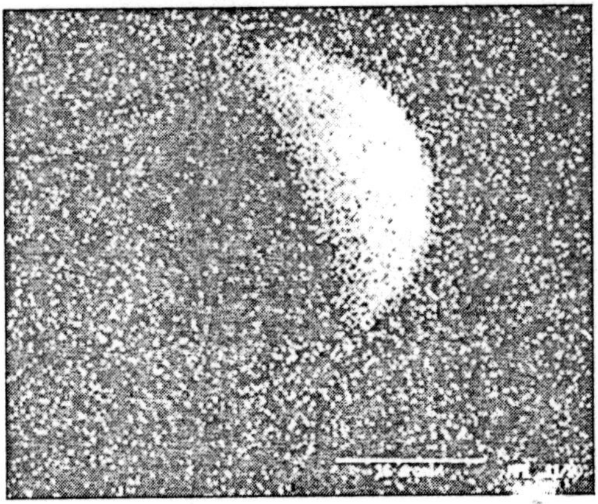

Fig. Rosat Image of X-ray Fluorescence of, and Occultation of the X-ray Background by, the Moon.

The use of X-rays for medical purposes (to develop into the field of radiation therapy) was pioneered by Major John Hall-Edwards in Birmingham, England. In 1908, he had to have his left arm amputated owing to the spread of X-ray dermatitis. The X-ray microscope was invented in the 1950s.

The Chandra X-ray Observatory, launched on July 23, 1999, has been allowing the exploration of the very violent processes in the universe which produce X-rays. Unlike visible light, which is a relatively stable view of the universe, the X-ray universe is unstable, it features stars being torn apart by black holes, galactic collisions, and novas, neutron stars that build up layers of plasma that then explode into space.

An X-ray laser device was proposed as part of the Reagan Administration's Strategic Defence Initiative in the 1980s, but the first and only test of the device (a sort of laser "blaster", or death ray, powered by a thermonuclear explosion) gave inconclusive results.

For technical and political reasons, the overall project (including the X-ray laser) was de-funded (though was later revived by the second Bush Administration as National Missile Defence using different technologies).

Chapter 11

Application of MRI

In clinical practice, MRI is used to distinguish pathologic tissue (such as a brain tumor) from normal tissue. One advantage of an MRI scan is that it is thought to be harmless to the patient. It uses strong magnetic fields and non-ionizing radiation in the radio frequency range. Compare this to CT scans and traditional X-rays which involve doses of ionizing radiation and may increase the risk of malignancy, especially in a fetus.

While CT provides good spatial resolution (the ability to distinguish two structures an arbitrarily small distance from each other as separate), MRI provides comparable resolution with far better contrast resolution (the ability to distinguish the differences between two arbitrarily similar but not identical tissues).

The basis of this ability is the complex library of *pulse sequences* that the modern medical MRI scanner includes, each of which is optimized to provide *image contrast* based on the chemical sensitivity of MRI. For example, with particular values of the *echo time* (TE) and the *repetition time* (TR), which are basic parameters of image acquisition, a sequence will take on the property of T2-weighting.

On a T2-weighted scan, fat-, water- and fluid-containing tissues are bright (most modern T2 sequences are actually *fast T2* sequences). Damaged tissue tends to develop edema, which makes a T2-weighted sequence sensitive for pathology, and generally able to distinguish pathologic tissue from normal tissue. With the addition of an additional radio frequency pulse and additional manipulation of the magnetic gradients, a T2-

weighted sequence can be converted to a FLAIR sequence, in which free water is now dark, but edematous tissues remain bright. This sequence in particular is currently the most sensitive way to evaluate the brain for demyelinating diseases, such as multiple sclerosis.

The typical MRI examination consists of 5-20 sequences, each of which are chosen to provide a particular type of information about the subject tissues. This information is then synthesized by the interpreting physician.

Specialized MRI Scans

Diffusion MRI

Diffusion MRI measures the diffusion of water molecules in biological tissues. In an isotropic medium (inside a glass of water for example) water molecules naturally move randomly according to Brownian motion. In biological tissues however, the diffusion may be anisotropic. For example a molecule inside the axon of a neuron has a low probability of crossing the myelin membrane.

Therefore the molecule will move principally along the axis of the neural fibre. If we know that molecules in a particular voxel diffuse principally in one direction we can make the assumption that the majority of the fibers in this area are going parallel to that direction.

The recent development of diffusion tensor imaging (DTI) enables diffusion to be measured in multiple directions and the fractional anisotropy in each direction to be calculated for each voxel. This enables researchers to make brain maps of fibre directions to examine the connectivity of different regions in the brain (using tractography) or to examine areas of neural degeneration and demyelination in diseases like Multiple Sclerosis.

Another application of diffusion MRI is diffusion-weighted imaging (DWI). Following an ischemic stroke, DWI is highly sensitive to the changes occurring in the lesion. It is speculated that increases in restriction (barriers) to water diffusion, as a result of cytotoxic edema (cellular swelling), is responsible for the increase in signal on a DWI scan. Other theories, including acute changes in cellular permeability and loss of energy-dependent (ATP) cytoplasmic streaming, have been proposed to explain the phenomena.

The DWI enhancement appears within 5-10 minutes of the onset of stroke symptoms (as compared with computed tomography, which often does not detect changes of acute infarct for up to 4-6 hours) and remains for up to two weeks. CT, due to its insensitivity to acute ischemia, is typically employed to rule out hemorrhagic stroke, which would entirely prevent the use of tissue plasminogen activator (tPA). Further, coupled with scans sensitized to cerebral perfusion, researchers can highlight regions of "perfusion/diffusion mismatch" that may indicate regions capable of salvage by reperfusion therapy.

Finally, it has been proposed that diffusion MRI may be able to detect minute changes in extracellular water diffusion and therefore could be used as a tool for fMRI. The nerve cell body enlarges when it conducts an action potential, hence restricting extracellular water molecules from diffusing naturally. Although this process works in theory, evidence is only moderately convincing. Like many other specialized applications, this technique is usually coupled with a fast image acquisition sequence, such as echo planar imaging sequence.

Magnetic Resonance Angiography

Magnetic resonance angiography (MRA) is used to generate pictures of the arteries in order to evaluate them for stenosis (abnormal narrowing) or aneurysms (vessel wall dilatations, at risk of rupture). MRA is often used to evaluate the arteries of the neck and brain, the thoracic and abdominal aorta, the renal arteries, and the legs (called a "run-off").

A variety of techniques can be used to generate the pictures, such as administration of a paramagnetic contrast agent (gadolinium) or using a technique known as "flow-related enhancement" (e.g. 2D and 3D time-of-flight sequences), where most of the signal on an image is due to blood which has recently moved into that plane. Magnetic resonance venography (MRV) is a similar procedure that is used to image veins. In this method the tissue is now excited inferiorly while signal is gathered in the plane immediately superior to the excitation plane, and thus imaging the venous blood which has recently moved from the excited plane.

Magnetic Resonance Spectroscopy

In vivo ('in the living organism') magnetic resonance spectroscopy (MRS), also known as MRSI (MRS imaging) and volume selective NMR spectroscopy, is a technique which combines the spatially-addressable nature of MRI with the spectroscopically-rich information obtainable from NMR.

That is to say, MRI allows one to study a particular region within an organism or sample, but gives relatively little information about the chemical or physical nature of that region (its chief value is in being able to distinguish the properties of that region, how much fat or water is present, relative to those of surrounding regions). MR spectroscopy, however, provides a wealth of information about other biological chemicals ('metabolites') within that region, as would an NMR spectrum of that region.

Functional MRI

Functional MRI (fMRI) measures signal changes in the brain that are due to changing neural activity. The brain is scanned at low resolution but at a rapid rate (typically once every 2-3 seconds). Increases in neural activity cause changes in the MR signal via T2* changes; this mechanism is referred to as the BOLD (blood-oxygen-level dependent) effect. Increased neural activity causes an increased demand for oxygen, and the vascular system actually overcompensates for this, increasing the amount of oxygenated hemoglobin relative to deoxygenated hemoglobin.

Because deoxygenated hemoglobin attenuates the MR signal, the vascular response leads to a signal increase that is related to the neural activity. The precise nature of the relationship between neural activity and the bold signal is a subject of current research. The bold effect also allows for the generation of high resolution 3D maps of the venous vasculature within neural tissue.

While BOLD signal is the most common method employed for neuroscience studies in human subjects, the flexible nature of MR imaging provides means to sensitize the signal to other aspects of the blood supply. Alternative techniques employ arterial spin labeling (ASL) or weight the MRI signal by cerebral blood flow (CBF) and cerebral blood volume (CBV).

The CBV method requires injection of a class of MRI contrast agents that are now in human clinical trials. Because this method has been shown to be far more sensitive than the BOLD technique in preclinical studies, it may potentially expand the role of fMRI in clinical applications. The CBF method provides more quantitative information than the BOLD signal, albeit at a significant loss of detection sensitivity.

Interventional MRI

The lack of harmful effects on the patient and the operator make MRI well-suited for "interventional radiology", where the images produced by a MRI scanner are used to guide minimally-invasive procedures.

Radiation Therapy Simulation

Because of MRI's superior imaging of soft tissues, it is now being utilized to specifically locate tumors within the body in preparation for radiation therapy treatments. For therapy simulation, a patient is placed in specific, reproducible, body position and scanned. The MRI system then computes the precise location, shape and orientation of the tumor mass, correcting for any spatial distortion inherent in the system. The patient is then marked or tattooed with points which, when combined with the specific body position, will permit precise triangulation for radiation therapy.

Current Density Imaging

Current density imaging (CDI) endeavors to use the phase

information from images to reconstruct current densities within a subject. Current density imaging works because electrical currents generate magnetic fields, which in turn affect the phase of the magnetic dipoles during an imaging sequence. To date no successful CDI has been performed using biological currents, but several studies have been published which involve applied currents through a pair of electrodes.

Magnetic Resonance Guided Focused Ultrasound

In MRgFUS therapy, ultrasound beams are focused on a tissue - guided and controlled using MR thermal imaging - and due to the significant energy deposition at the focus, temperature within the tissue rises to more than 65°C, completely destroying it. This technology can achieve precise "ablation" of diseased tissue. MR imaging provides a three-dimensional view of the target tissue, allowing for precise focusing of ultrasound energy.

The MR imaging provides quantitative, real-time, thermal images of the treated area. This allows the physician to ensure that the temperature generated during each cycle of ultrasound energy is sufficient to cause thermal ablation within the desired tissue and if not, to adapt the parameters to ensure effective treatment.

Multinuclear Imaging

Hydrogen is the most frequently imaged nucleus in MRI because it is present in biological tissues in great abundance. However, any nucleus which has a net nuclear spin could potentially be imaged with MRI. Such nuclei include helium-3, carbon-13, fluorine-19, oxygen-17, sodium-23, phosphorus-31 and xenon-129. ^{23}Na and ^{31}P are naturally abundant in the body, so can be imaged directly.

Gaseous isotopes such as ^{3}He or ^{129}Xe must be hyperpolarized and then inhaled as their nuclear density is too low to yield a useful signal under normal conditions. ^{17}O, ^{13}C and ^{19}F can be administered in sufficient quantities in liquid form (e.g. ^{17}O-water, ^{13}C-glucose solutions or perfluorocarbons) that hyperpolarization is not a necessity.

Multinuclear imaging is primarily a research technique at present. However, potential applications include functional imaging and imaging of organs poorly seen on ^1H MRI (e.g. lungs and bones) or as alternative contrast agents. Inhaled hyperpolarized ^3He can be used to image the distribution of air spaces within the lungs.

Injectable solutions containing ^{13}C or stabilized bubbles of hyperpolarized ^{129}Xe have been studied as contrast agents for angiography and perfusion imaging. ^{31}P can potentially provide information on bone density and structure, as well as functional imaging of the brain.

Experimental MRI Techniques

Currently there is active research in several new MRI technologies like magnetization transfer MRI (MT-MRI), diffusion tensor MRI (DT-MRI), and proton MR spectroscopy, plus recent research in to Dendrimer-enhanced MRI as a diagnostic and prognostic biomarker of sepsis-induced acute renal failure.

SAFETY

Implants and foreign bodies: Pacemakers are generally considered an absolute contraindication towards MRI scanning, though highly specialized protocols have been developed to permit scanning of select pacing devices. Several cases of arrhythmia or death have been reported in patients with pacemakers who have undergone MRI scanning without appropriate precautions.

Other electronic implants have varying contraindications, depending upon scanner technology, implant properties, scanning protocols and anatomy being imaged. Though pacemakers receive significant attention, it should also be noted that many other forms of medical or biostimulation implants may be contraindicated for MRI scans.

These may include Vagus nerve stimulators, implantable cardioverter-defibrillators (ICD), loop recorders, insulin pumps, cochlear implants, deep brain stimulators and many others. Medical device patients should always present complete information (manufacturer, model, serial number and date of implantation) about all implants to both the referring physician and to the radiologist or technologist before entering the room for the MRI scan.

While these implants pose a current problem, scientist and manufacturers are working on improved designs which will further minimize the risks that MRI scans pose to medical device operations.

One such development in the works is a nano coating for implants intended to screen them from the radio frequency waves, helping to make MRI exams available to patients currently prohibited from receiving them.

Ferromagnetic foreign bodies (e.g. shell fragments), or metallic implants (e.g. surgical prostheses, aneurysm clips) are also potential risks, and safety aspects need to be considered on an individual basis. Interaction of the magnetic and radio frequency fields with such objects can lead to: trauma due to movement of the object in the magnetic field, thermal injury from radio-frequency induction heating of the object, or failure of an implanted device.

These issues are especially problematic when dealing with the eye. Most MRI centers require an orbital x-ray be performed on anyone who suspects they may have small metal fragments in their eyes, perhaps from a previous accident, something not uncommon in metalworking.

Because of its non-ferromagnetic nature and poor electrical conductivity, titanium and its alloys are useful for long term implants and surgical instruments intended for use in image-guided surgery. In particular, not only is titanium safe from movement from the magnetic field, but artifacts around the implant are less frequent and less severe than with more ferromagnetic materials e.g. stainless steel.

Artifacts from metal frequently appear as regions of empty space around the implant - frequently called 'black-hole artifact' e.g. a 3mm titanium alloy coronary stent may appear as a 5mm diameter region of empty space on MRI, whereas around a stainless steel stent, the artifact may extend for 10-20 mm or more.

Projectile or Missile Effect

As a result of the very high strength of the magnetic field needed to produce scans (frequently up to 60,000 times the earth's own magnetic field effects), there are several incidental safety issues addressed in MRI facilities. Missile-effect accidents, where ferromagnetic objects are attracted to the centre of the magnet, have resulted in injury and death.

A video simulation of a fatal projectile effect accident illustrates the extreme power that contemporary MRI equipment can exert on ferromagnetic objects. In order to help reduce the risks of projectile accidents, ferrous objects and devices are typically prohibited in proximity to the MRI scanner, with non ferro-magnetic versions of many tools and devices typically retained by the scanning facility.

Patients undergoing MRI examinations are required to remove all metallic objects, often by changing into a gown or 'scrubs'. The magnetic field and the associated risk of missile-effect accidents remains a permanent hazard — as superconductive MRI magnets retain their magnetic field, even in the event of a power outage.

Radio Frequency Energy

A powerful radio transmitter is needed for excitation of proton spins. This can heat the body significantly, with the risk of hyperthermia in patients, particularly the obese or patients with thermoregulation disorders. Several countries have issued restrictions on the maximum specific absorption rate that a scanner may produce.

Peripheral Nerve Stimulation

The rapid switching (on and off) of the magnetic field gradients needed for imaging is capable of causing nerve stimulation. Volunteers report a twitching sensation when exposed to rapidly switched fields, particularly in their extremities. The reason the peripheral nerves are stimulated is that the changing field increases with distance from the centre of the gradient coils (which more or less coincides with the centre of the magnet).

Note however that when imaging the head, the heart is far off-centre and induction of even a tiny current into the heart must be avoided at all costs. Although PNR was not a problem for the slow, weak gradients used in the early days of MRI, the strong, rapidly-switched gradients used in techniques such as EPI, fMRI, diffusion MRI, etc. are indeed capable of inducing PNR.

American and European regulatory agencies insist that manufacturers stay below specified dB/dt limits (dB/dt is the change in field per unit time) or else prove (via clinical studies) that no PNR is induced for any imaging sequence. As a result of dB/dt limitation software and/or hardware, commercial MRI systems cannot use the full rated power of their gradient amplifiers.

Acoustic Noise

Loud noises and vibrations are produced by forces resulting from rapidly switched magnetic gradients interacting with the main magnetic field, in turn causing minute expansions and contractions of the coil itself.

This is most marked with high-field machines and rapid-imaging techniques in which sound intensity can reach 130 dB (equivalent to a jet engine at take-off). Appropriate use of ear protection is essential for anyone inside the MRI scanner room during the examination.

Cryogens

As described above in 'Scanner Construction And Operation', many MRI scanners rely on cryogenic liquids to enable superconducting capabilities of the electromagnetic coils within. Though the cryogenic liquids most frequently used are non-toxic, their physical properties present specific hazards.

An emergency shut-down of a superconducting electromagnet, an operation known as "quenching", involves the rapid boiling of liquid helium from the device. If the rapidly expanding helium cannot be dissipated though an external vent, sometimes referred to as 'quench pipe', it may be released into the scanner room where it may cause displacement of the oxygen and present a risk of asphyxiation. Liquid helium, the most commonly used cryogen in MRI, undergoes near explosive expansion as it changes from liquid to a gaseous state.

Rooms built in support of superconducting MRI equipment should be equipped with pressure relief mechanisms and an exhaust fan, in addition to the required quench pipe. Since a quench results in rapid loss of all cryogens in the magnet, recommissioning the magnet is extremely expensive and time-consuming.

Spontaneous quenches are uncommon, but can occur at any time. Quenches may also be triggered by equipment malfunction, improper cryogen fill technique, contaminates inside the cryostat, or extreme magnetic or vibrational disturbances.

Contrast Agents

The most frequently used intravenous contrast agents are based on chelates of gadolinium. In general, these agents have proved safer than the iodinated contrast agents used in X-ray radiography or CT. Anaphylactoid reactions are rare occurring in approx 0.03-0.1 per cent. Of particular interest is the lower incidence of nephrotoxicity, compared with iodinated agents, when given at usual doses—this has made contrast-enhanced MRI scanning an option for patients with renal impairment, who would otherwise not be able to undergo contrast-enhanced CT.

Although gadolinium agents have proved useful for patients with renal impairment, in patients with severe renal failure requiring dialysis there is a risk of a rare but serious illness, nephrogenic systemic fibrosis, that may be linked to the use of certain gadolinium-containing agents: the most frequently linked is gadodiamide, but other agents have been linked too.

Although a causal link has not been definitively established, current guidelines in the United States are that dialysis patients should only receive gadolinium agents where essential, and that dialysis should be performed as soon as possible after the scan is complete, in order to remove the agent from the body promptly. In Europe where more gadolinium-containing agents are available, a classification of agents according to potential risks has been released.

Pregnancy

No harmful effects of MRI on the fetus have been demonstrated. In particular, MRI avoids the use of ionizing radiation, to which the fetus is particularly sensitive. However, as a precaution, current guidelines recommend that pregnant women undergo MRI only when essential. This is particularly the case during the first trimester of pregnancy, as organogenesis takes place during this period.

The concerns in pregnancy are the same as for MRI in general, but the fetus may be more sensitive to the effects—particularly to heating and to noise. However, one additional concern is the use of contrast agents; gadolinium compounds are known to cross the placenta and enter the fetal bloodstream, and it is recommended that their use be avoided.

Despite these concerns, MRI is rapidly growing in importance as a way of diagnosing and monitoring congenital defects of the fetus because it can provide more diagnostic information than ultrasound and it lacks the ionizing radiation of CT. MRI without contrast is the imaging mode of choice for pre-surgical, in-utero diagnosis and evaluation of fetal tumors, primarily teratomas, facilitating open fetal surgery, other fetal interventions, and planning for procedures (such as the EXIT procedure) to safely deliver and treat babies whose defects would otherwise be fatal.

Claustrophobia and Discomfort

Due to the construction of MRI scanners, they are potentially unpleasant in which to lie. The part of the body being imaged needs to lie at the centre of the magnet (which is often a long, narrow tube). Because scan times may be long (perhaps one hour), people with even mild claustrophobia are often unable to tolerate an MRI scan without management.

Management may include:

- Advance preparation
  - Visiting the scanner
  - Visualization techniques
  - Chemical sedation
  - General anesthesia
- Coping while inside the scanner
  - Holding a "panic button"
  - Listening to music on headphones or watching a movie with a Head-mounted display while in the machine
- Modified scanner designs
  - Upright MRIs
  - Open-bore design scanners.

Though open MRIs have increased in popularity as of late, they produce inferior scan quality because they operate at lower magnetic fields than closed MRIs. For babies and children, chemical sedation or general anesthesia is the norm. These MRI subjects are too young to be instructed to hold still during the scanning session.

Obese patients and pregnant women may find the MRI machine to be a tight fit, and even some claustrophobics may find the experience intolerable without sedation. Pregnant women may also have difficulty lying on their backs for an hour or more without moving.

The noise associated with the operation of an MRI scanner (especially the audible noise associated with the gradient pulses applied to the subject) can also exacerbate the discomfort associated with the procedure.

The European Physical Agents Directive

The European Physical Agents (Electromagnetic Fields) Directive is European legislation that has been adopted in European legislature. By 2008 each individual state within the European Union must include this directive in its own law.

The directive applies to occupational exposure to electromagnetic fields (not medical exposure) and was intended to limit workers' acute exposure to strong electromagnetic fields, as may be found near electricity substations, radio or television transmitters or industrial equipment.

However, the regulations impact significantly on MRI, with separate sections of the regulations limiting exposure to static magnetic fields, changing magnetic fields and radio frequency energy. Field strength limits are given which may not be exceeded for any period of time. An employer may commit a criminal offense by allowing a worker to exceed an exposure limit if that is how the Directive is implemented in a particular Member State.

The Directive is based on the international consensus of established effects of exposure to electromagnetic fields, and in particular the advice of the European Commissions's advisor, the International Commission on Non-Ionizing Radiation Protection (ICNIRP).

The aims of the Directive, and the ICNIRP guidelines upon which it is based, are to prevent exposure to potentially harmful fields. The actual limits in the Directive are very similar to the limits advised by the Institute of Electrical and Electronics Engineers, with the exception of the frequencies produced by the gradient coils, where the IEEE limits are significantly higher.

Many Member States of the EU already have either specific EMF regulations or (as in the UK) a general requirement under workplace health and safety legislation to protect workers against electromagnetic fields.

In almost all cases the existing regulations are aligned with the ICNIRP limits so that the Directive should, in theory, have little impact on any employer already meeting their legal responsibilities.

The introduction of the Directive has brought to light an existing potential issue with occupational exposures to MRI fields. There are at present very few data on the number or types of MRI practice that might lead to exposures in excess of the levels of the Directive. There is a justifiable concern amongst MRI practitioners that if the Directive were to be enforced more vigorously than existing legislation, the use of

MRI might be restricted, or working practices of MRI personnel might have to change. In the initial draft a limit of static field strength to 2 T was given.

This has since been removed from the regulations, and whilst it is unlikely to be restored as it was without a strong justification, some restriction on static fields may be reintroduced after the matter has been considered more fully by ICNIRP. The effect of such a limit might be to restrict the installation, operation and maintenance of MRI scanners with magnets of 2 T and stronger.

As the increase in field strength has been instrumental in developing higher resolution and higher performance scanners, this would be a significant step back. This is why it is unlikely to happen without strong justification.

Individual government agencies and the European Commission have now formed a working group to examine the implications on MRI and to try to address the issue of occupational exposures to electromagnetic fields from MRI.

Chapter 12

Artificial Fusion

NUCLEAR FUSION

In nuclear physics and nuclear chemistry, nuclear fusion is the process by which multiple like-charged atomic nuclei join together to form a heavier nucleus. It is accompanied by the release or absorption of energy. Iron and nickel nuclei have the largest binding energies per nucleon of all nuclei. The fusion of two nuclei with lower mass than iron generally releases energy while the fusion of nuclei heavier than iron absorbs energy; vice-versa for the reverse process, nuclear fission.

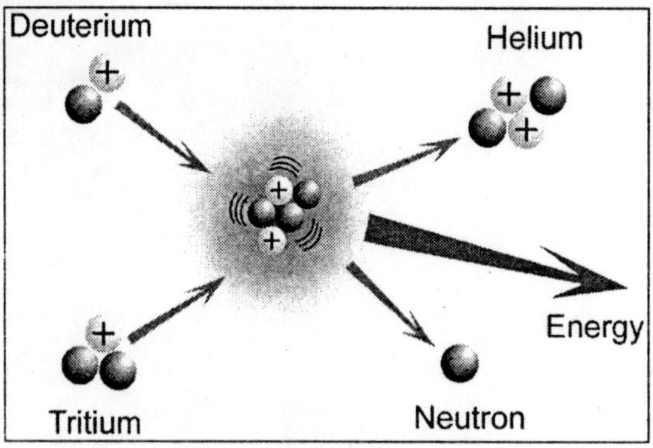

Fig. Nuclear fusion

Nuclear fusion occurs naturally in stars. Artificial fusion in human enterprises has also been achieved, although not yet completely controlled. Building upon the nuclear transmutation experiments of Ernest Rutherford done a few years earlier, fusion of light nuclei (hydrogen isotopes) was first observed by Mark Oliphant in 1932, and the steps of the main cycle of nuclear fusion in stars were subsequently worked out by Hans Bethe throughout the remainder of that decade.

Research into fusion for military purposes began in the early 1940s, as part of the Manhattan Project, but was not successful until 1952. Research into controlled fusion for civilian purposes began in the 1950s, and continues to this day.

Fusion reactions power the stars and produce all but the lightest elements in a process called nucleosynthesis. While the fusion of lighter elements in stars releases energy, production of the heavier elements absorbs energy. When the fusion reaction is a sustained uncontrolled chain, it can result in a thermonuclear explosion, such as that generated by a hydrogen bomb. Reactions which are not self-sustaining can still release considerable energy, as well as large numbers of neutrons.

Research into controlled fusion, with the aim of producing fusion power for the production of electricity, has been conducted for over 50 years. It has been accompanied by extreme scientific and technological difficulties, but resulted in steady progress.

As of the present, break-even (self-sustaining) controlled fusion reaction have been demonstrated in a few tokamak-type reactors around the world and resulted in producing workable design of the reactor which will deliver ten times more fusion energy than the amount of energy needed to heat up to force nuclei to fuse, even those of the lightest element, hydrogen. This is because all nuclei have a positive charge (due to their protons), and as like charges repel, nuclei strongly resist being put too close together.

Accelerated to high speeds (that is, heated to thermonuclear temperatures), they can overcome this electromagnetic repulsion and get close enough for the attractive nuclear force to be sufficiently strong to achieve fusion. The fusion of lighter nuclei, creating a heavier nucleus and a free neutron, will generally release more energy than it took to force them together-an exothermic process that can produce self-sustaining reactions. The energy released in most nuclear reactions is much larger than that in chemical reactions, because the binding energy that holds a nucleus together is far greater than the energy that holds electrons to a nucleus. For example, the ionization energy gained by adding an electron to a hydrogen nucleus is 13.6 electron volts - less than one-millionth of the 17 MeV released in the D-T (deuterium-tritium) reaction shown to the top right.

Fusion reactions have an energy density many times greater than nuclear fission — i.e., per unit of mass the reactions produce far greater energies, even though *individual* fission reactions are generally much more energetic than *individual* fusion reactions-which are themselves millions of times more energetic than chemical reactions. Only the direct conversion of mass into energy, such as with collision of matter and antimatter, is more energetic per unit of mass than nuclear fusion.

NUCLEAR FORCE

A substantial energy barrier of electrostatic forces must be overcome before fusion can occur. At large distances two naked nuclei repel one another because of the repulsive electrostatic force between their positively charged protons. If two nuclei can be brought close enough together, however, the electrostatic repulsion can be overcome by the attractive nuclear force which is stronger at close distances.

When a nucleon such as a proton or neutron is added to a nucleus, the nuclear force attracts it to other nucleons, but primarily to its immediate neighbours due to the short range of the force. The nucleons in the interior of a nucleus have more neighboring nucleons than those on the surface.

Since smaller nuclei have a larger surface area-to-volume ratio, the binding energy per nucleon due to the strong force generally increases with the size of the nucleus but approaches a limiting value corresponding to that of nucleus with a diameter of about four nucleons.

The electrostatic force, on the other hand, is an inverse-square force, so a proton added to a nucleus will feel an electrostatic repulsion from *all* the other protons in the nucleus. The electrostatic energy per nucleon due to the electrostatic force thus increases without limit as nuclei get larger.

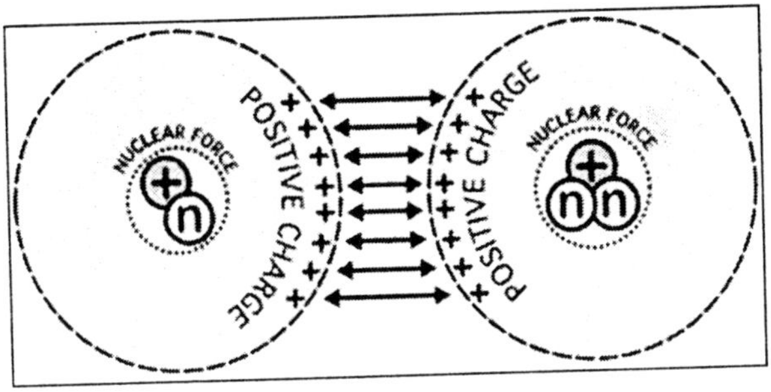

Fig. The Repulsion (Electrostatic Force) between Positively-Charged Nuclei is very weak over Long Distances, but Much Stronger at Shorter Distances. As Duch, the Main Technical Difficulty for Fusion is Getting the Nuclei Close Enough to Fuse.

The net result of these opposing forces is that the binding energy per nucleon generally increases with increasing size, up to the elements iron and nickel, and then decreases for heavier nuclei.

Eventually, the binding energy becomes negative and very heavy nuclei (all with more than 208 nucleons, corresponding to a diameter of about 6 nucleons) are not stable. The four most tightly bound nuclei, in decreasing order of binding energy, are ^{62}Ni, ^{58}Fe, ^{56}Fe, and ^{60}Ni. Even though the nickel isotope,Ni, is more stable, the iron isotope ^{56}Fe is an order of magnitude more common. This is due to a greater disintegration rate for ^{62}Ni in the interior of stars driven by photon absorption.

A notable exception to this general trend is the helium-4 nucleus, whose binding energy is higher than that of lithium, the next heavier element. The Pauli exclusion principle provides an explanation for this exceptional behaviour—it says that because protons and neutrons are fermions, they cannot exist in exactly the same state.

Each proton or neutron energy state in a nucleus can accommodate both a spin up particle and a spin down particle. Helium-4 has an anomalously large binding energy because its nucleus consists of two protons and two neutrons; so all four of its nucleons can be in the ground state. Any additional nucleons would have to go into higher energy states.

The situation is similar if two nuclei are brought together. As they approach each other, all the protons in one nucleus repel all the protons in the other. Not until the two nuclei actually come in contact can the strong nuclear force take over. Consequently, even when the final energy state is lower, there is a large energy barrier that must first be overcome. It is called the Coulomb barrier.

The Coulomb barrier is smallest for isotopes of hydrogen—they contain only a single positive charge in the nucleus. A bi-proton is not stable, so neutrons must also be involved, ideally in such a way that a helium nucleus, with its extremely tight binding, is one of the products.

Using deuterium-tritium fuel, the resulting energy barrier is about 0.01 MeV. In comparison, the energy needed to remove an electron from hydrogen is 13.6 eV, about 750 times less energy. The (intermediate) result of the fusion is an unstable ^{5}He nucleus, which immediately ejects a neutron with 14.1 MeV. The recoil energy of the remaining ^{4}He nucleus is 3.5 MeV, so the total energy liberated is 17.6 MeV. This is many times more than what was needed to overcome the energy barrier.

If the energy to initiate the reaction comes from accelerating one of the nuclei, the process is called *beam-target* fusion; if both nuclei are accelerated, it is *beam-beam* fusion. If the nuclei are part of a plasma near thermal equilibrium, one speaks of *thermonuclear* fusion.

Temperature is a measure of the average kinetic energy of particles, so by heating the nuclei they will gain energy and eventually have enough to overcome this 0.01 MeV. Converting the units between electronvolts and kelvins shows that the barrier would be overcome at a temperature in excess of 120 million kelvins, obviously a very high temperature.

There are two effects that lower the actual temperature needed. One is the fact that temperature is the *average* kinetic energy, implying that some nuclei at this temperature would actually have much higher energy than 0.01 MeV, while others would be much lower. It is the nuclei in the high-energy tail of the velocity distribution that account for most of the fusion reactions.

The other effect is quantum tunneling. The nuclei do not actually have to have enough energy to overcome the Coulomb barrier completely. If they have nearly enough energy, they can tunnel through the remaining barrier. For this reason fuel at lower temperatures will still undergo fusion events, at a lower rate.

The fusion reaction rate increases rapidly with temperature until it maximizes and then gradually drops off. The DT rate peaks at a lower temperature (about 70 keV, or 800 million kelvins) and at a higher value than other reactions commonly considered for fusion energy.

The reaction cross section ó is a measure of the probability of a fusion reaction as a function of the relative velocity of the two reactant nuclei. If the reactants have a distribution of velocities, e.g. a thermal distribution with thermonuclear fusion, then it is useful to perform an average over the distributions of the product of cross section and velocity. The reaction rate (fusions per volume per time) is <óv> times the product of the reactant number densities:

$$f = n_1 n_2 \langle \sigma \upsilon \rangle$$

If a species of nuclei is reacting with itself, such as the DD reaction, then the product $n_1 n_2$ must be replaced by $(1/2)n^2$. $\langle \sigma \upsilon \rangle$ increases from virtually zero at room temperatures up to meaningful magnitudes at temperatures of 10 – 100 keV. At these temperatures, well above typical ionization energies (13.6 eV in the hydrogen case), the fusion reactants exist in a plasma state. The significance of <σv> as a function of temperature in a device with a particular energy confinement time is found by considering the Lawson criterion.

GRAVITATIONAL CONFINEMENT

One force capable of confining the fuel well enough to satisfy the Lawson criterion is gravity. The mass needed, however, is so great that gravitational confinement is only found in stars (the smallest of which are brown dwarfs). Even if the more reactive fuel deuterium were used, a mass greater than that of the planet Jupiter would be needed.

MAGNETIC CONFINEMENT

The charged ions of fusion fuel and magnetic field lines and the fuel is therefore trapped along the field lines. A variety of magnetic configurations exist, including the toroidal geometries of tokamaks and stellarators and open ended mirror confinement systems.

INERTIAL CONFINEMENT

A third confinement principle is to apply a rapid pulse of energy to a large part of the surface of a pellet of fusion fuel, causing it to simultaneously "implode" and heat to very high pressure and temperature. If the fuel is dense enough and hot enough, the fusion reaction rate will be high enough to burn a significant fraction of the fuel before it has dissipated.

To achieve these extreme conditions, the initially cold fuel must be explosively compressed. Inertial confinement is used in the hydrogen bomb, where the driver is x-rays created by a fission bomb. Inertial confinement is also attempted in "controlled" nuclear fusion, where the driver is a laser, ion, or electron beam, or a Z-pinch.

Another method is to use conventional high explosive material to compress a fuel to fusion conditions.The UTIAS explosive-driven-implosion facility was used to produce stable, centered and focused hemispherical implosions to generate neutrons from D-D reactions.

The simplest and most direct method proved to be in a predetonated stoichiometric mixture of deuterium-oxygen.The other successful method was using a miniature Voitenko compressor,where a plane diaphragm was driven by the implosion wave into a secondary small spherical cavity that contained pure deuterium gas at one atmosphere. Some confinement principles have been investigated, such as muon-catalyzed fusion, the Farnsworth-Hirsch fusor and Polywell (inertial electrostatic confinement), and bubble fusion.

PRODUCTION METHODS

A variety of methods are known to effect nuclear fusion. Some are "cold" in the strict sense that no part of the material is hot (except for the reaction products), some are "cold" in the limited sense that the bulk of the material is at a relatively low temperature and pressure but the reactants are not, and some are "hot" fusion methods that create macroscopic regions of very high temperature and pressure.

Locally Cold Fusion

Muon-catalyzed fusion is a well-established and reproducible fusion process that occurs at ordinary temperatures. It was studied in detail by Steven Jones in the early 1980s.

It has not been reported to produce net energy. Net energy production from this reaction is not believed to be possible because of the energy required to create muons, their 2.2 μs half-life, and the chance that a muon will bind to the new alpha particle and thus stop catalyzing fusion.

Generally Cold, Locally Hot Fusion

Accelerator based light-ion fusion. Using particle accelerators it is possible to achieve particle kinetic energies sufficient to induce many light ion fusion reactions. Accelerating light ions is relatively easy, cheap, and can be done in an efficient manner - all it takes is a vacuum tube, a pair of electrodes, and a high-voltage transformer; fusion can be observed with as little as 10 kilovolt between electrodes.

The key problem with accelerator-based fusion (and with cold targets in general) is that fusion cross sections are many orders of magnitude lower than Coulomb interaction cross sections. Therefore vast majority of ions ends up expending their energy on bremsstrahlung and ionization of atoms of the target.

Devices referred to as sealed-tube neutron generators are particularly relevant to this discussion. These small devices are miniature particle accelerators filled with deuterium and tritium gas in an arrangement which allows ions of these nuclei to be accelerated against hydride targets, also containing deuterium and tritium, where fusion takes place.

Hundreds of neutron generators are produced annually for use in the petroleum industry where they are used in measurement equipment for locating and mapping oil reserves.

Despite periodic reports in the popular press by scientists claiming to have invented "table-top" fusion machines, neutron generators have been around for half a century. The sizes of these devices vary but the smallest instruments are often packaged in sizes smaller than a loaf of bread. These devices do not produce a net power output.

- In sonoluminescence, acoustic shock waves create temporary bubbles that collapse shortly after creation, producing very high temperatures and pressures. In 2002, Rusi P. Taleyarkhan reported the possibility that bubble fusion occurs in those collapsing bubbles (aka sonofusion). As of 2005, experiments to determine whether fusion is occurring give conflicting results.
  If fusion is occurring, it is because the local temperature and pressure are sufficiently high to produce hot fusion. In an episode of Horizon, on BBC television, results were presented showing that, although temperatures were reached which could initiate fusion on a large scale, no fusion was occurring, and inaccuracies in the measuring system were the cause of anomalous results.
- The Farnsworth-Hirsch Fusor is a tabletop device in which fusion occurs. This fusion comes from high effective temperatures produced by electrostatic acceleration of ions. The device can be built inexpensively, but it too is unable to produce a net power output.
- Antimatter-initialized fusion uses small amounts of antimatter to trigger a tiny fusion explosion. This has been studied primarily in the context of making nuclear pulse propulsion feasible. This is not near

becoming a practical power source, due to the cost of manufacturing antimatter alone.

- Pyroelectric fusion was reported in April 2005 by a team at UCLA. The scientists used a pyroelectric crystal heated from "34 to 7°C ("30 to 45°F), combined with a tungsten needle to produce an electric field of about 25 gigavolts per meter to ionize and accelerate deuterium nuclei into an erbium deuteride target.

  Though the energy of the deuterium ions generated by the crystal has not been directly measured, the authors used 100 keV (a temperature of about 10^9 K) as an estimate in their modeling.

  At these energy levels, two deuterium nuclei can fuse together to produce a helium-3 nucleus, a 2.45 MeV neutron and bremsstrahlung. Although it makes a useful neutron generator, the apparatus is not intended for power generation since it requires far more energy than it produces.

HOT FUSION

In "standard" "hot" fusion, the fuel reaches tremendous temperature and pressure inside a fusion reactor or nuclear weapon. The methods in the second group are examples of non-equilibrium systems, in which very high temperatures and pressures are produced in a relatively small region adjacent to material of much lower temperature.

In his doctoral thesis for MIT, Todd Rider did a theoretical study of all quasineutral, isotropic, non-equilibrium fusion systems. He demonstrated that all such systems will leak energy at a rapid rate due to bremsstrahlung radiation produced when electrons in the plasma hit other electrons or ions at a cooler temperature and suddenly decelerate.

The problem is not as pronounced in a hot plasma because the range of temperatures, and thus the magnitude of the deceleration, is much lower. Note that Rider's work does not apply to non-neutral and/or anisotropic non-equilibrium plasmas.

IMPORTANT REACTIONS

Astrophysical Reaction Chains

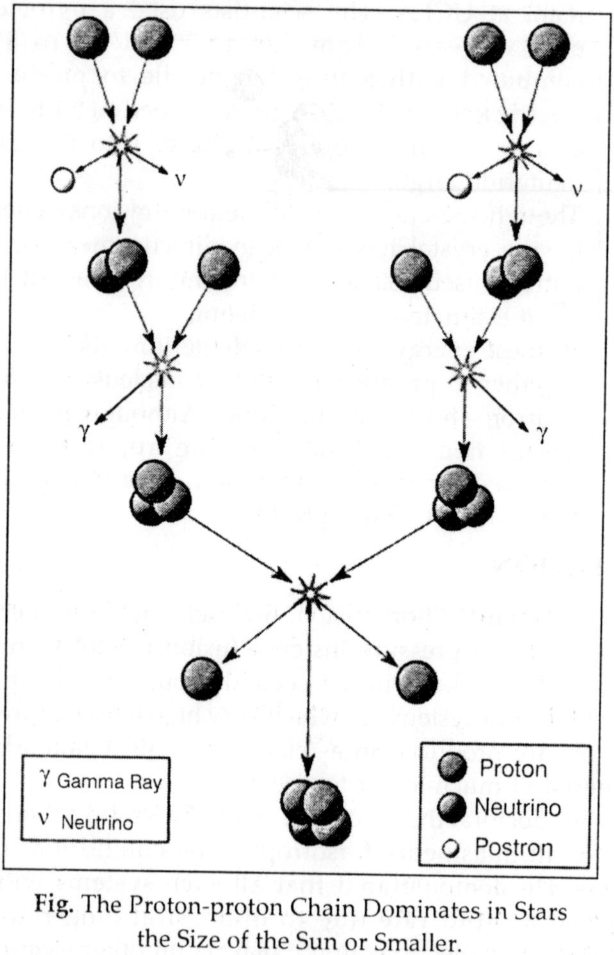

Fig. The Proton-proton Chain Dominates in Stars the Size of the Sun or Smaller.

The most important fusion process in nature is that which powers the stars. The net result is the fusion of four protons into one alpha particle, with the release of two positrons, two neutrinos (which changes two of the protons into neutrons), and energy, but several individual reactions are involved, depending on the mass of the star. For stars the size of the sun or smaller, the proton-proton chain dominates. In heavier stars, the CNO cycle is more important. Both types of processes are responsible for the creation of new elements as part of stellar nucleosynthesis.

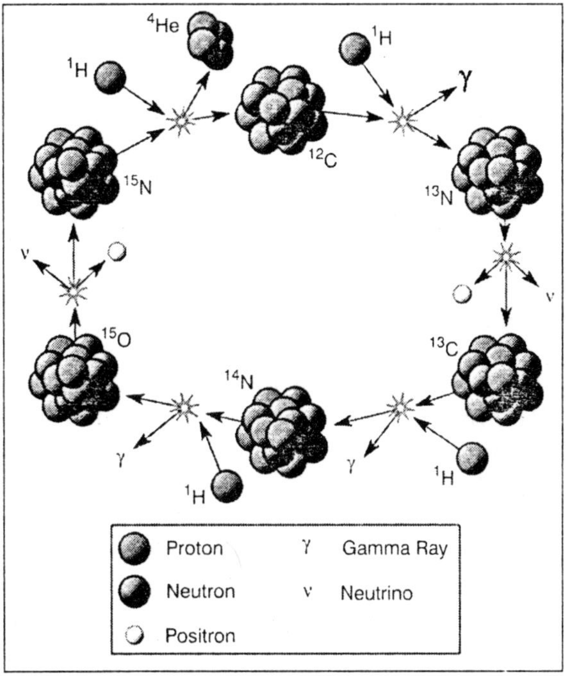

Fig. The CNO Cycle Dominates in Stars Heavier than the Sun.

At the temperatures and densities in stellar cores the rates of fusion reactions are notoriously slow. For example, at solar core temperature ($T \approx 15$ MK) and density (160 g/cm^3), the energy release rate is only 276 μW/cm^3—about a quarter of the volumetric rate at which a resting human body generates heat. Thus, reproduction of stellar core conditions in a lab for nuclear fusion power production is completely impractical.

Because nuclear reaction rates strongly depend on temperature ($\exp(-E/kT)$), then in order to achieve reasonable rates of energy production in terrestrial fusion reactors 10–100 times higher temperatures (compared to stellar interiors) are required $T \approx$ 0.1–1.0 GK.

CRITERIA AND CANDIDATES FOR TERRESTRIAL REACTIONS

In man-made fusion, the primary fuel is not constrained to be protons and higher temperatures can be used, so reactions with larger cross-sections are chosen. This implies a lower Lawson criterion, and therefore less startup effort.

Another concern is the production of neutrons, which activate the reactor structure radiologically, but also have the advantages of allowing volumetric extraction of the fusion energy and tritium breeding. Reactions that release no neutrons are referred to as *aneutronic.*

In order to be useful as a source of energy, a fusion reaction must satisfy several criteria. It must

- Be exothermic: This may be obvious, but it limits the reactants to the low Z (number of protons) side of the curve of binding energy. It also makes helium ^{4}He the most common product because of its extraordinarily tight binding, although ^{3}He and ^{3}H also show up.
- Involve low Z nuclei: This is because the electrostatic repulsion must be overcome before the nuclei are close enough to fuse.
- Have two reactants: At anything less than stellar densities, three body collisions are too improbable. It should be noted that in inertial confinement, both stellar densities and temperatures are exceeded to compensate for the shortcomings of the third parameter of the Lawson criterion, ICF's very short confinement time.
- Have two or more products: This allows simultaneous conservation of energy and momentum without relying on the electromagnetic force.
- Conserve both protons and neutrons: The cross sections for the weak interaction are too small.

Few reactions meet these criteria. The following are those with the largest cross sections:

For reactions with two products, the energy is divided between them in inverse proportion to their masses, as shown.

In most reactions with three products, the distribution of energy varies. For reactions that can result in more than one set of products, the branching ratios are given. Some reaction candidates can be eliminated at once.

No.	Reactants	Products	Energy released	Branching
(1)	$^{2}_{1}D + ^{3}_{1}T$	$\rightarrow ^{4}_{2}He$ (3.5 MeV) + n^0 (14.1 MeV)		
(2i)	$^{2}_{1}D + ^{2}_{1}D$	$\rightarrow ^{3}_{1}T$ (1.01 MeV) + p^+ (3.02 MeV)		50%
(2ii)		$\rightarrow ^{3}_{2}He$ (0.82 MeV) + n^0 (2.45 MeV)		50%
(3)	$^{2}_{1}D + ^{3}_{2}He$	$\rightarrow ^{4}_{2}He$ (3.6 MeV) + p^+ (14.7 MeV)		
(4)	$^{3}_{1}T + ^{3}_{1}T$	$\rightarrow ^{4}_{2}He + 2n^0$	+ 11.3 MeV	
(5)	$^{3}_{2}He + ^{3}_{2}He$	$\rightarrow ^{4}_{2}He + 2p^+$	+ 12.9 MeV	
(6i)	$^{3}_{2}He + ^{3}_{1}T$	$\rightarrow ^{4}_{2}He + p^+ + n^0$	+ 12.1 MeV	51%
(6ii)		$\rightarrow ^{4}_{2}He$ (4.8 MeV) + $^{2}_{1}D$ (9.5 MeV)		43%
(6iii)		$\rightarrow ^{4}_{2}He$ (0.5 MeV) + n^0 (1.9 MeV) + p^+ (11.9 MeV)		6%
(7i)	$^{2}_{1}D + ^{6}_{3}Li$	$\rightarrow 2\,^{4}_{2}He$ + 22.4 MeV		
(7ii)		$\rightarrow ^{3}_{2}He + ^{4}_{2}He + n^0$	+ 2.56 MeV	
(7iii)		$\rightarrow ^{7}_{3}Li + p^+$	+ 5.0 MeV	
(7iv)		$\rightarrow ^{7}_{4}Be + n^0$	+ 3.4 MeV	
(8)	$p^+ + ^{6}_{3}Li$	$\rightarrow ^{4}_{2}He$ (1.7 MeV) + $^{3}_{2}He$ (2.3 MeV)		
(9)	$^{3}_{2}He + ^{6}_{3}Li$	$\rightarrow 2\,^{4}_{2}He + p^+$	+ 16.9 MeV	
(10)	$p^+ + ^{11}_{5}B$	$\rightarrow 3\,^{4}_{2}He$	+ 8.7 MeV	

The D-^{6}Li reaction has no advantage compared to $p^+ - ^{11}_{5}B$ because it is roughly as difficult to burn but produces substantially more neutrons through $^{2}_{1}D - ^{2}_{1}$ side reactions.

There is also a $p^+ - ^{7}_{3}Li$ reaction, but the cross section is far too low, except possibly when $T_i > 1$ MeV, but at such high temperatures an endothermic, direct neutron-producing reaction also becomes very significant.

Finally there is also a $p^+ - ^{9}_{4}Be$ reaction, which is not only difficult to burn, but $^{9}_{4}Be$ can be easily induced to split into two alpha particles and a neutron. In addition to the fusion reactions, the following reactions with neutrons are important in order to "breed" tritium in "dry" fusion bombs and some proposed fusion reactors:

$$n^0 + {}^6_3\text{Li} \rightarrow {}^3_1\text{T} + {}^4_2\text{He}$$

$$n^0 + {}^7_3\text{Li} \rightarrow {}^3_1\text{T} + {}^4_2\text{He} + n^0$$

To evaluate the usefulness of these reactions, in addition to the reactants, the products, and the energy released, one needs to know something about the cross section. Any given fusion device will have a maximum plasma pressure that it can sustain, and an economical device will always operate near this maximum. Given this pressure, the largest fusion output is obtained when the temperature is chosen so that <óv>/T^2 is a maximum.

This is also the temperature at which the value of the triple product *nTô* required for ignition is a minimum, since that required value is inversely proportional to <óv>/T^2. (A plasma is "ignited" if the fusion reactions produce enough power to maintain the temperature without external heating.) This optimum temperature and the value of <óv>/T^2 at that temperature is given for a few of these reactions in the following table.

Nucleosynthesis		
Fuel	T[keV]	$<\sigma v>/T^2[m^2/s/keV^2]$
${}^2_1\text{D} - {}^3_1\text{T}$	13.6	1.24×10^{-24}
${}^2_1\text{D} - {}^2_1\text{D}$	15	1.28×10^{-26}
${}^2_1\text{D} - {}^3_2\text{He}$	58	2.24×10^{-26}
$p^+ - {}^3_3\text{Li}$	66	1.46×10^{-27}
$p^+ - {}^{11}_5\text{B}$	123	3.01×10^{-27}

Note that many of the reactions form chains. For instance, a reactor fueled with $^{3}_{1}T$ and $^{3}_{2}He$ will create some $^{2}_{1}D$, which is then possible to use in the $^{2}_{1}D - ^{3}_{2}He$ reaction if the energies are "right". An elegant idea is to combine the reactions. The $^{3}_{2}He$ from reaction can react with $^{6}_{3}Li$ in reaction before completely thermalizing.

This produces an energetic proton which in turn undergoes reaction before thermalizing. A detailed analysis shows that this idea will not really work well, but it is a good example of a case where the usual assumption of a Maxwellian plasma is not appropriate.

NEUTRONICITY

Any of the reactions above can in principle be the basis of fusion power production. In addition to the temperature and cross section discussed above, we must consider the total energy of the fusion products E_{fus}, the energy of the charged fusion products E_{ch}, and the atomic number Z of the non-hydrogenic reactant.

Specification of the $^{2}_{1}D - ^{2}_{1}D$ reaction entails some difficulties, though. To begin with, one must average over the two branches (2) and (3). More difficult is to decide how to treat the $^{3}_{1}T$ and $^{3}_{2}He$ products. $^{3}_{1}T$ burns so well in a deuterium plasma that it is almost impossible to extract from the plasma. The $^{2}_{1}D - ^{3}_{2}He$ reaction is optimized at a much higher temperature, so the burnup at the optimum $^{2}_{1}D - ^{2}_{1}D$ temperature may be low, so it seems reasonable to assume the $^{3}_{1}T$ but not the $^{3}_{2}He$ gets burned up and adds its energy to the net reaction. Thus we will count the $^{2}_{1}D - ^{2}_{1}D$ fusion energy as E_{fus} = (4.03+17.6+3.27)/2 = 12.5 MeV and the energy in charged particles as E_{ch} = (4.03+3.5+0.82)/2 = 4.2 MeV. Another unique aspect of the ${}_1^2D - {}_1^2D$ reaction is that there is only one reactant, which must be taken into account when calculating the reaction rate.

With this choice, we tabulate parameters for four of the most important reactions.

fuel	Z	E_{fus} [MeV]	E_{ch} [MeV]	neutronicity
${}_1^2D$-${}_1^3T$	1	17.6	3.5	0.80
${}_1^2D$-${}_1^2D$	1	12.5	4.2	0.66
${}_1^2D$-${}_2^3He$	2	18.3	18.3	~0.05
p^+-${}_5^{11}B$	5	8.7	8.7	~0.001

The last column is the neutronicity of the reaction, the fraction of the fusion energy released as neutrons. This is an important indicator of the magnitude of the problems associated with neutrons like radiation damage, biological shielding, remote handling, and safety.

For the first two reactions it is calculated as $(E_{fus}-E_{ch})/E_{fus}$. For the last two reactions, where this calculation would give zero, the values quoted are rough estimates based on side reactions that produce neutrons in a plasma in thermal equilibrium.

Of course, the reactants should also be mixed in the optimal proportions. This is the case when each reactant ion plus its associated electrons accounts for half the pressure. Assuming that the total pressure is fixed, this means that density of the non-hydrogenic ion is smaller than that of the hydrogenic ion by a factor 2/(Z+1).

Therefore the rate for these reactions is reduced by the same factor, on top of any differences in the values of <óv>/T².

On the other hand, because the ${}_1^2D - {}_1^2D$ reaction has only one reactant, the rate is twice as high as if the fuel were divided between two hydrogenic species.

Thus there is a "penalty" of (2/(Z+1)) for non-hydrogenic

fuels arising from the fact that they require more electrons, which take up pressure without participating in the fusion reaction. (It is usually a good assumption that the electron temperature will be nearly equal to the ion temperature. Some authors, however discuss the possibility that the electrons could be maintained substantially colder than the ions. In such a case, known as a "hot ion mode", the "penalty" would not apply.) There is at the same time a "bonus" of a factor 2 for ${}_1^2D - {}_1^2D$ because each ion can react with any of the other ions, not just a fraction of them.

Knowing that an atomic mass unit u = 1.66x10^{27}Kg

2.014102+3.016050 = 4.002603 + 1.008665 + E

5.030152u + 5.011268u + E

the difference in mass is 0.01884u, when we put this equation into $E=MC^2$

$E=(3.134744\text{x}10^{-29})\text{x}(3\text{x}10^{8}) = E=2.82\text{x}10^{-12}J$

and knowing that 1ev is equal to 1.602x10^{-19}J

therefore 17610921.35ev is produced which is 17.6Mev, 14.1Mev is given to the electron and 3.6Mev is given to He_4

The maximum value of <óv>/T^2 is taken from a previous table. The "penalty/bonus" factor is that related to a non-hydrogenic reactant or a single-species reaction. The values in the column "reactivity" are found by dividing 1.24×10^{-24} by the product of the second and third columns. It indicates the factor by which the other reactions occur more slowly than the ${}_1^2D - {}_1^3T$ reaction under comparable conditions. The column "Lawson criterion" weights these results with E_{ch} and gives an indication of how much more difficult it is to achieve ignition with these reactions, relative to the difficulty for the ${}_1^2D - {}_1^3T$ reaction. The last column is labeled "power density" and weights the practical reactivity with E_{fus}. It indicates how much lower the fusion power density of the other reactions is compared to the ${}_1^2D - {}_1^3T$ reaction and can be considered a measure of the economic potential.

Bremsstrahlung Losses in Quasineutral, Isotropic Plasmas

The ions undergoing fusion in many systems will essentially never occur alone but will be mixed with electrons that in aggregate neutralize the ions' bulk electrical charge and form a plasma. The electrons will generally have a temperature comparable to or greater than that of the ions, so they will collide with the ions and emit x-ray radiation of 10-30 keV energy (Bremsstrahlung).

The Sun and stars are opaque to x-rays, but essentially any terrestrial fusion reactor will be optically thin for x-rays of this energy range. X-rays are difficult to reflect but they are effectively absorbed (and converted into heat) in less than mm thickness of stainless steel (which is part of a reactor's shield). The ratio of fusion power produced to x-ray radiation lost to walls is an important figure of merit.

This ratio is generally maximized at a much higher temperature than that which maximizes the power density. The following table shows the rough optimum temperature and the power ratio at that temperature for several reactions.

fuel	T_i (keV)	$P_{fusion}/P_{Bremsstrahlung}$
${}^{2}_{1}D$-${}^{3}_{1}T$	50	140
${}^{2}_{1}D$-${}^{2}_{1}D$	500	2.9
${}^{2}_{1}D$-${}^{3}_{2}He$	100	5.3
${}^{3}_{2}He$-${}^{3}_{2}He$	1000	0.72
p^{+}-${}^{6}_{3}Li$	800	0.21
p^{+}-${}^{11}_{5}B$	300	0.57

The actual ratios of fusion to Bremsstrahlung power will likely be significantly lower for several reasons. For one, the calculation assumes that the energy of the fusion products is transmitted completely to the fuel ions, which then lose energy to the electrons by collisions, which in turn lose energy by Bremsstrahlung. However because the fusion products move much faster than the fuel ions, they will give up a significant fraction of their energy directly to the electrons. Secondly, the plasma is assumed to be composed purely of fuel ions. In practice, there will be a significant proportion of impurity ions, which will lower the ratio.

n particular, the fusion products themselves *must* remain in the plasma until they have given up their energy, and *will* remain some time after that in any proposed confinement scheme. Finally, all channels of energy loss other than Bremsstrahlung have been neglected.

The last two factors are related. On theoretical and experimental grounds, particle and energy confinement seem to be closely related. In a confinement scheme that does a good job of retaining energy, fusion products will build up. If the fusion products are efficiently ejected, then energy confinement will be poor, too.

The temperatures maximizing the fusion power compared to the Bremsstrahlung are in every case higher than the temperature that maximizes the power density and minimizes the required value of the fusion triple product. This will not change the optimum operating point for ${}_1^2D-{}_1^3T$ very much because the Bremsstrahlung fraction is low, but it will push the other fuels into regimes where the power density relative to ${}_1^2D-{}_1^3T$ is even lower and the required confinement even more difficult to achieve. ${}_1^2D-{}_1^2D$ and ${}_1^2D-{}_2^3He$ For, Bremsstrahlung losses will be a serious, possibly prohibitive problem. For ${}_2^3He-{}_2^3He, p^+-{}_3^6Li$ and $p^+-{}_5^{11}B$ the Bremsstrahlung losses appear to make a fusion reactor using these fuels with a quasineutral, anisotropic plasma impossible.

Chapter 13

Fission of Heavy Elements

NUCLEAR FISSION

In nuclear physics and nuclear chemistry, nuclear fission is a nuclear reaction in which the nucleus of an atom splits into smaller parts, often producing free neutrons and lighter nuclei, which may eventually produce photons (in the form of gamma rays). Fission of heavy elements is an exothermic reaction which can release large amounts of energy both as electromagnetic radiation and as kinetic energy of the fragments (heating the bulk material where fission takes place). Fission is a form of nuclear transmutation because the resulting fragments are not the same element as the original atom.

Nuclear fission produces energy for nuclear power and to drive the explosion of nuclear weapons. Both uses are made possible because certain substances called nuclear fuels undergo fission when struck by free neutrons and in turn generate neutrons when they break apart. This makes possible a self-sustaining chain reaction that releases energy at a controlled rate in a nuclear reactor or at a very rapid uncontrolled rate in a nuclear weapon.

The amount of free energy contained in nuclear fuel is millions of times the amount of free energy contained in a similar mass of chemical fuel such as gasoline, making nuclear fission a very tempting source of energy; however, the products of nuclear fission are radioactive and remain so for significant amounts of time, giving rise to a nuclear waste problem. Concerns over nuclear waste accumulation and over the destructive potential of nuclear weapons may counterbalance the desirable qualities of fission as an energy source, and give rise to ongoing political debate over nuclear power.

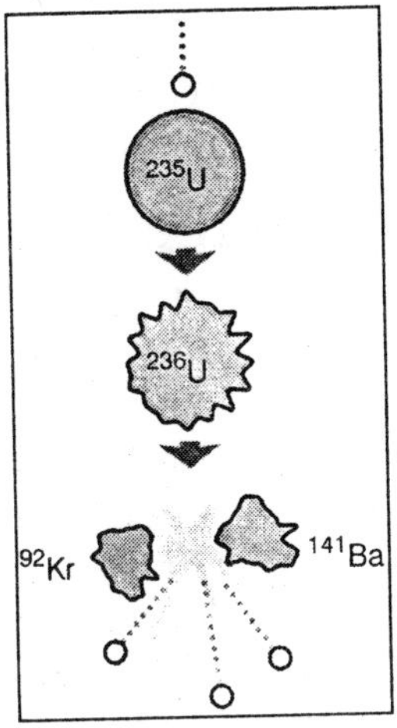

Fig. An Induced Nuclear Fission Event. A Slow-moving Neutron is Absorbed by the Nucleus of a Uranium-235 Atom, which in Turn Splits into Fast-moving Lighter Elements (Fission Products) and Free Neutrons.

Nuclear fission differs from other forms of radioactive decay in that it can be harnessed and controlled via a chain reaction: free neutrons released by each fission event can trigger yet more events, which in turn release more neutrons and cause more fissions. Chemical isotopes that can sustain a fission chain reaction are called nuclear fuels, and are said to be fissile. The most common nuclear fuels are ^{235}U (the isotope of uranium with an atomic mass of 235 and of use in nuclear reactors) and ^{239}Pu (the isotope of plutonium with an atomic mass of 239).

These fuels break apart into a bimodal range of chemical elements with atomic masses centering near 95 and 135 u (fission products). Most nuclear fuels undergo spontaneous fission only very slowly, decaying mainly via an alpha/beta decay chain over periods of millennia to eons. In a nuclear reactor or nuclear weapon, most fission events are induced by bombardment with another particle such as a neutron. Typical fission events release about two hundred million eV of energy for each fission event. By contrast, most chemical oxidation reactions (such as burning coal or TNT) release at most a few eV per event, so nuclear fuel contains at least ten million times more usable energy than does chemical fuel.

The energy of nuclear fission is released as kinetic energy of the fission products and fragments, and as electromagnetic radiation in the form of gamma rays; in a nuclear reactor, the energy is converted to heat as the particles and gamma rays collide with the atoms that make up the reactor and its working fluid, usually water or occasionally heavy water.

When uranium decays into two other atoms, there is produced an energy of ~200 MeV. Of this energy, 168 MeV is kinetic energy of uranium parts (atoms), which fly away with a speed of c/30. Also there is emitted an average of 2.5 neutrons with kinetic energy of ~2 MeV each. Finally, the fission reaction emits ~30 MeV gamma photon.

Neutron and gamma rays emitted by fragments erase information about fission process itself, making difficult to study dynamics from saddle to scission point, where the fragments are formed and nuclear interaction dispels. Nevertheless there are few fission events for which no neutron nor gamma are emitted. These events are part of the so called cold fission.

In fission there is a preference to yield fragments with even proton numbers, which is called odd-even effect on fragments charge distribution. However, no odd-even effect is observed on fragment mass number distribution. This result is attributed to nucleon pair breaking in fission process.

Nuclear fission of heavy elements produces energy because the specific binding energy (binding energy per mass) of intermediate-mass nuclei with atomic numbers and atomic masses close to ^{61}Ni and ^{56}Fe is greater than the specific binding energy of very heavy nuclei, so that energy is released when heavy nuclei are broken apart. The total rest masses of the fission products (Mp) from a single reaction is less than the mass of the original fuel nucleus (M). The excess mass

Äm = M – Mp is the invariant mass of the energy that is released as photons (gamma rays) and kinetic energy of the fission fragments, according to the mass-energy equivalence formula $E = mc^2$. In nuclear fission events the nuclei may break into any combination of lighter nuclei, but the most common event is not fission to equal mass nuclei of about mass 120; the most common event (depending on isotope and process) is a slightly unequal fission in which one daughter nucleus has a mass of about 90 to 100 u and the other the remaining 130 to 140 u.

Unequal fissions are energetically more favorable because this allows one product to be closer to the energetic minimum near mass 60 u (only a quarter of the average fissionable mass), while the other nucleus with mass 135 u is still not far out of the range of the most tightly bound nuclei (another statement of this, is that the atomic binding energy curve is slightly steeper to the left of mass 120 u than to the right of it).

The variation in specific binding energy with atomic number is due to the interplay of the two fundamental forces acting on the component nucleons (protons and neutrons) that make up the nucleus. Nuclei are bound by an attractive strong nuclear force between nucleons, which overcomes the electrostatic repulsion between protons.

However, the strong nuclear force acts only over extremely short ranges, since it follows a Yukawa potential. For this reason large nuclei are less tightly bound per unit mass than small nuclei, and breaking a very large nucleus into two or more intermediate-sized nuclei releases energy. Because of the short range of the strong binding force, large nuclei must contain proportionally more neutrons than do light elements, which are most stable with a 1–1 ratio of protons and neutrons.

Extra neutrons stabilize heavy elements because they add to strong-force binding without adding to proton-proton repulsion. Fission products have, on average, about the same ratio of neutrons and protons as their parent nucleus, and are therefore usually unstable because they have proportionally too many neutrons compared to stable isotopes of similar mass. This is the fundamental cause of the problem of radioactive high level waste from nuclear reactors. Fission products tend to be beta emitters, emitting fast-moving electrons to conserve electric charge as excess neutrons convert to protons inside the nucleus of the fission product atoms.

The most common nuclear fuels, ^{235}U and ^{239}Pu, are not major radiologic hazards by themselves: ^{235}U has a half-life of approximately 700 million years, and although ^{239}Pu has a half-life of only about 24,000 years, it is a pure alpha particle emitter and hence not particularly dangerous unless ingested.

Once a fuel element has been used, the remaining fuel material is intimately mixed with highly radioactive fission products that emit energetic beta particles and gamma rays. Some fission products have half-lives as short as seconds; others have half-lives of tens of thousands of years, requiring long-term storage in facilities such as Yucca Mountain until the fission products decay into non-radioactive stable isotopes.

NUCLEAR CHAIN REACTIONS

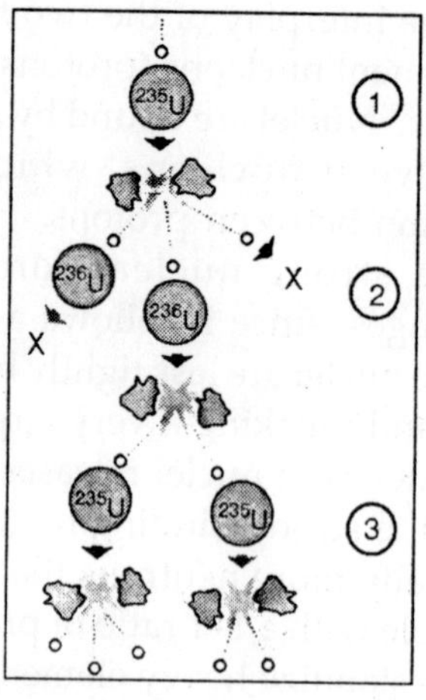

Fig. A Schematic Nuclear Fission Chain Reaction.

- A uranium-235 atom absorbs a neutron and fissions into two new atoms (fission fragments), releasing three new neutrons and some binding energy. 2. One of those neutrons is absorbed by an atom of uranium-238 and does not continue the reaction.

Another neutron is simply lost and does not collide with anything, also not continuing the reaction. However one neutron does collide with an atom of uranium-235, which then fissions and releases two neutrons and some binding energy. 3. Both of those neutrons collide with uranium-235 atoms, each of which fissions and releases between one and three neutrons, which can then continue the reaction.

Many heavy elements, such as uranium, thorium, and plutonium, undergo both spontaneous fission, a form of radioactive decay and *induced fission*, a form of nuclear reaction. Elemental isotopes that undergo induced fission when struck by a free neutron are called fissionable; isotopes that undergo fission when struck by a thermal, slow moving neutron are also called fissile.

A few particularly fissile and readily obtainable isotopes (notably ^{235}U and ^{239}Pu) are called nuclear fuels because they can sustain a chain reaction and can be obtained in large enough quantities to be useful. All fissionable and fissile isotopes undergo a small amount of spontaneous fission which releases a few free neutrons into any sample of nuclear fuel. Such neutrons would escape rapidly from the fuel and become a free neutron, with a half-life of about 15 minutes before they decayed to protons and beta particles.

However, neutrons almost invariably impact and are absorbed by other nuclei in the vicinity long before this happens (newly-created fission neutrons are moving at about 7 per cent of the speed of light, and even moderated neutrons are moving at about 8 times the speed of sound). Some neutrons will impact fuel nuclei and induce further fissions, releasing yet more neutrons.

If enough nuclear fuel is assembled into one place, or if the escaping neutrons are sufficiently contained, then these freshly generated neutrons outnumber the neutrons that escape from the assembly, and a *sustained nuclear chain reaction* will take place.

An assembly that supports a sustained nuclear chain reaction is called a critical assembly or, if the assembly is almost entirely made of a nuclear fuel, a critical mass. The word "critical" refers to a cusp in the behaviour of the differential equation that governs the number of free neutrons present in the fuel:

If less than a critical mass is present, then the amount of neutrons is determined by radioactive decay, but if a critical mass or more is present, then the amount of neutrons is controlled instead by the physics of the chain reaction. The actual mass of a *critical mass* of nuclear fuel depends strongly on the geometry and surrounding materials.

Not all fissionable isotopes can sustain a chain reaction. For example, ^{238}U, the most abundant form of uranium, is fissionable but not fissile: it undergoes induced fission when impacted by an energetic neutron with over 1 MeV of kinetic energy. But too few of the neutrons produced by ^{238}U fission are energetic enough to induce further fissions in ^{238}U, so no chain reaction is possible with this isotope.

Instead, bombarding ^{238}U with slow neutrons causes it to absorb them (becoming ^{239}U) and decay by beta emission to ^{239}Np which then decays again by the same process to ^{239}Pu; that process is used to manufacture ^{239}Pu in breeder reactors, but does not contribute to a neutron chain reaction.

Fissionable, non-fissile isotopes can be used as fission energy source even without a chain reaction. Bombarding ^{238}U with fast neutrons induces fissions, releasing energy as long as the external neutron source is present.That effect is used to augment the energy released by modern thermonuclear weapons, by jacketing the weapon with ^{238}U to react with neutrons released by nuclear fusion at the centre of the device.

FISSION REACTORS

Critical fission reactors are the most common type of nuclear reactor. In a critical fission reactor, neutrons produced by fission of fuel atoms are used to induce yet more fissions, to sustain a controllable amount of energy release. Devices that produce engineered but non-self-sustaining fission reactions are *subcritical fission reactors*. Such devices use radioactive decay or particle accelerators to trigger fissions. Critical fission reactors are built for three primary purposes, which typically involve different engineering trade-offs to take advantage of either the heat or the neutrons produced by the fission chain reaction:

- *Power reactors* are intended to produce heat for nuclear power, either as part of a generating station or a local power system such as a nuclear submarine.
- *Research reactors* are intended to produce neutrons and/or activate radioactive sources for scientific, medical, engineering, or other research purposes.
- *Breeder reactors* are intended to produce nuclear fuels in bulk from more abundant isotopes. The better known fast breeder reactor makes ^{239}Pu (a nuclear fuel) from the naturally very abundant ^{238}U (not a nuclear fuel). Thermal breeder reactors previously tested using ^{232}Th continue to be studied and developed.

While, in principle, all fission reactors can act in all three capacities, in practice the tasks lead to conflicting engineering goals and most reactors have been built with only one of the above tasks in mind. (There are several early counter-examples, such as the Hanford N reactor, now decommissioned). Power reactors generally convert the kinetic energy of fission products into heat, which is used to heat a working fluid and drive a heat engine that generates mechanical or electrical power.

The working fluid is usually water with a steam turbine, but some designs use other materials such as gaseous helium. Research reactors produce neutrons that are used in various ways, with the heat of fission being treated as an unavoidable waste product. Breeder reactors are a specialized form of research reactor, with the caveat that the sample being irradiated is usually the fuel itself, a mixture of ^{238}U and ^{235}U.

FISSION BOMBS

One class of nuclear weapon, a *fission bomb* (not to be confused with the *fusion bomb*), otherwise known as an *atomic bomb* or *atom bomb*, is a fission reactor designed to liberate as much energy as possible as rapidly as possible, before the released energy causes the reactor to explode (and the chain reaction to stop). Development of nuclear weapons was the motivation behind early research into nuclear fission: the Manhattan Project of the U.S. military during World War II carried out most of the early scientific work on fission chain reactions, culminating in the Little Boy and Fat Man and Trinity bombs that were exploded over test sites, the cities Hiroshima, and Nagasaki, Japan in August 1945.

Even the first fission bombs were thousands of times more explosive than a comparable mass of chemical explosive. For example, Little Boy weighed a total of about four tons (of which 60 kg was nuclear fuel) and was 11 feet (3.4 m) long; it also yielded an explosion equivalent to about 15 kilotons of TNT, destroying a large part of the city of Hiroshima.

Modern nuclear weapons (which include a thermonuclear *fusion* as well as one or more fission stages) are literally hundreds of times more energetic for their weight than the first pure fission atomic bombs, so that a modern single missile warhead bomb weighing less than 1/8th as much as Little Boy has a yield of 475,000 tons of TNT, and could bring destruction to 10 times the city area.

While the fundamental physics of the fission chain reaction in a nuclear weapon is similar to the physics of a controlled nuclear reactor, the two types of device must be engineered quite differently.

It is impossible to convert a nuclear reactor to cause a true nuclear explosion, or for a nuclear reactor to explode the way a nuclear explosive does, (though partial fuel meltdowns and steam explosions have occurred), and difficult to extract useful power from a nuclear explosive (though at least one rocket propulsion system, Project Orion, was intended to work by exploding fission bombs behind a massively padded vehicle).

The strategic importance of nuclear weapons is a major reason why the technology of nuclear fission is politically sensitive. Viable fission bomb designs are, arguably, within the capabilities of bright undergraduates being incredibly simple, but nuclear fuel to realise the designs is thought to be difficult to obtain being rare.

Chapter 14

Fusion Power

Fusion power is power generated by nuclear fusion reactions. In this kind of reaction, two light atomic nuclei fuse together to form a heavier nucleus and in doing so, release energy. In a more general sense, the term can also refer to the production of net usable power from a fusion source, similar to the usage of the term "steam power."

Most design studies for fusion power plants involve using the fusion reactions to create heat, which is then used to operate a steam turbine, similar to most coal-fired power stations as well as fission-driven nuclear power stations.

FUEL CYCLE

The basic concept behind any fusion reaction is to bring two or more atoms close enough together so that the strong nuclear force in their nuclei will pull them together into one larger atom. If two light nuclei fuse, they will generally form a single nucleus with a slightly smaller mass than the sum of their original masses.

The difference in mass is released as energy according to Einstein's mass-energy equivalence formula $E = mc^2$. If the input atoms are sufficiently massive, the resulting fusion product will be heavier than the reactants, in which case the reaction requires an external source of energy.

The dividing line between "light" and "heavy" is iron-56. Above this atomic mass, energy will generally be released by nuclear fission reactions; below it, by fusion. Fusion between the atoms is opposed by their shared electrical charge, specifically the net positive charge of the nuclei. In order to overcome this electrostatic force, or "Coulomb barrier", some external source of energy must be supplied. The easiest way to do this is to heat the atoms, which has the side effect of stripping the electrons from the atoms and leaving them as bare nuclei.

In most experiments the nuclei and electrons are left in a fluid known as a plasma. The temperatures required to provide the nuclei with enough energy to overcome their repulsion is a function of the total charge, so hydrogen, which has the smallest nuclear charge therefore reacts at the lowest temperature.

Helium has an extremely low mass per nucleon and therefore is energetically favoured as a fusion product. As a consequence, most fusion reactions combine isotopes of hydrogen ("protium", deuterium, or tritium) to form isotopes of helium (^3He or ^4He).

Perhaps the three most widely considered fuel cycles are based on the D-T, D-D, and p-^{11}B reactions. Other fuel cycles (D-^3He and ^3He-^3He) would require a supply of ^3He, either from other nuclear reactions or from extraterrestrial sources, such as the surface of the moon or the atmospheres of the gas giant planets. The details of the calculations comparing these reactions can be found here.

D-T Fuel Cycle

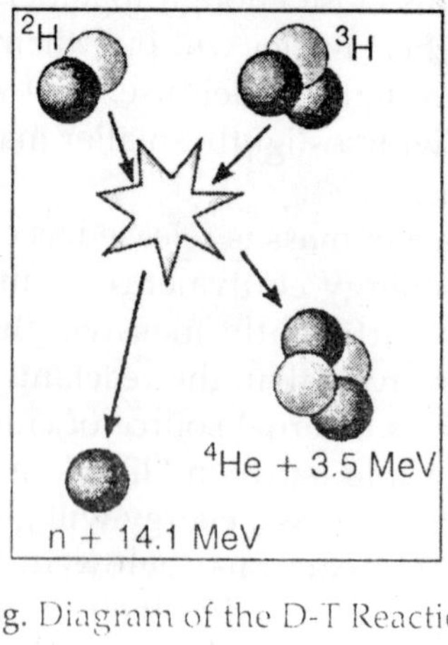

Fig. Diagram of the D-T Reaction

The easiest (according to the Lawson criterion) and most immediately promising nuclear reaction to be used for fusion power is:

$$D + T \rightarrow {}^{4}He + n$$

Deuterium is a naturally occurring isotope of hydrogen and as such is universally available. The large mass ratio of the hydrogen isotopes makes the separation rather easy compared to the difficult uranium enrichment process. Tritium is also an isotope of hydrogen, but it occurs naturally in only negligible amounts due to its radioactive half-life of 12.32 years. Consequently, the deuterium-tritium fuel cycle requires the breeding of tritium from lithium using one of the following reactions:

$$n + {}^{6}Li \rightarrow T + {}^{4}He$$

$$n + {}^{7}Li \rightarrow T + {}^{4}He + n$$

The reactant neutron is supplied by the D-T fusion reaction shown above, the one which also produces the useful energy. The reaction with ${}^{6}Li$ is exothermic, providing a small energy gain for the reactor. The reaction with ${}^{7}Li$ is endothermic but does not consume the neutron. At least some ${}^{7}Li$ reactions are required to replace the neutrons lost by reactions with other elements. Most reactor designs use the naturally occurring mix of lithium isotopes.

The supply of lithium is more limited than that of deuterium, but still large enough to supply the world's energy demand for thousands of years. Several drawbacks are commonly attributed to D-T fusion power:

- It produces substantial amounts of neutrons that result in induced radioactivity within the reactor structure.
- Only about 20 per cent of the fusion energy yield appears in the form of charged particles (the rest neutrons), which limits the extent to which direct energy conversion techniques might be applied.
- The use of D-T fusion power depends on lithium resources, which are less abundant than deuterium resources.
- It requires the handling of the radioisotope tritium.

Similar to hydrogen, tritium is difficult to contain and may leak from reactors in some quantity. Some estimates suggest that this would represent a fairly large environmental release of radioactivity.

The neutron flux expected in a commercial D-T fusion reactor is about 100 times that of current fission power reactors, posing problems for material design. Design of suitable materials is under way but their actual use in a reactor is not proposed until the generation after ITER.

After a single series of D-T tests at JET, the largest fusion reactor yet to use this fuel, the vacuum vessel was sufficiently radioactive that remote handling needed to be used for the year following the tests.

On the other hand, the volumetric deposition of neutron power can also be seen as an advantage. If all the power of a fusion reactor had to be transported by conduction through the surface enclosing the plasma, it would be very difficult to find materials and a construction that would survive, and it would probably entail a relatively poor efficiency.

D-D Fuel Cycle

Though more difficult to facilitate than the deuterium-tritium reaction, fusion can also be achieved through the reaction of deuterium with itself. This reaction has two branches that occur with nearly equal probability:

$$\begin{aligned} D + D &\rightarrow T + p \\ &\rightarrow {}^3He + n \end{aligned}$$

The optimum temperature for this reaction is 15 keV, only slightly higher than the optimum for the D-T reaction. The first branch does not produce neutrons, but it does produce tritium, so that a D-D reactor will not be completely tritium-free, even though it does not require an input of tritium or lithium.

Most of the tritium produced will be burned before leaving the reactor, which reduces the tritium handling required, but also means that more neutrons are produced and that some of these are very energetic. The neutron from the second branch has an energy of only 2.45 MeV, whereas the neutron from the D-T reaction has an energy of 14.1 MeV, resulting in a wider range of isotope production and material damage. Assuming complete tritium burn-up, the reduction in the fraction of fusion energy carried by neutrons is only about 18 per cent, so that the primary advantage of the D-D fuel cycle is that tritium breeding is not required.

Other advantages are independence from limitations of lithium resources and a somewhat softer neutron spectrum. The price to pay compared to D-T is that the energy confinement (at a given pressure) must be 30 times better and the power produced (at a given pressure and volume) is 68 times less.

p-^{11}B Fuel Cycle

If aneutronic fusion is the goal, then the most promising candidate may be the proton-boron reaction:

$$p + {}^{11}B \rightarrow 3\,{}^{4}He$$

Under reasonable assumptions, side reactions will result in about 0.1 per cent of the fusion power being carried by neutrons. At 123 keV, the optimum temperature for this reaction is nearly ten times higher than that for the pure hydrogen reactions, the energy confinement must be 500 times better than that required for the D-T reaction, and the power density will be 2500 times lower than for D-T.

Since the confinement properties of conventional approaches to fusion such as the tokamak and laser pellet fusion are marginal, most proposals for aneutronic fusion are based on radically different confinement concepts.

Research for Fusion Reaction

The idea of using human-initiated fusion reactions was first made practical for military purposes, in nuclear weapons. In a hydrogen bomb, the energy released by a fission weapon is used to compress and heat fusion fuel, beginning a fusion reaction which can release a very large amount of energy.

The first fusion-based weapons released some 500 times more energy than early fission weapons. Civilian applications, in which explosive energy production must be replaced by a controlled production, are still being developed. Although it took less than ten years to go from military applications to civilian fission energy production, it has been very different in the fusion energy field; more than fifty years have already passed without any commercial fusion energy production plant coming into operation.

Magnetic Approach

Registration of the first patent related to a fusion reactor by the United Kingdom Atomic Energy Authority, the inventors being Sir George Paget Thomson and Moses Blackman, dates back to 1946. Some basic principles used in the ITER experiment are described in this patent: toroidal vacuum chamber, magnetic confinement, and radio frequency plasma heating. The U.S. fusion programme began in 1951 when Lyman Spitzer began work on a stellarator under the code name Project Matterhorn.His work led to the creation of the Princeton Plasma Physics Laboratory, where magnetically confined plasmas are still studied. The stellarator concept fell out of favour for several decades afterwards, plagued by poor confinement issues, but recent advances in computer technology have led to a significant resurgence in interest in these devices.

A wide variety of other magnetic geometries were also experimented with, notably with the magnetic mirror. These systems also suffered from similar problems when higher performance versions were constructed. A new approach was outlined in the theoretical works fulfilled in 1950–1951 by I.E. Tamm and A.D. Sakharov in Soviet Union, which laid the foundations of the tokamak. Experimental research of these systems started in 1956 in Kurchatov Institute, Moscow by a group of Soviet scientists lead by Lev Artsimovich.

The group constructed the first tokamaks, the most successful of them being T-3 and its larger version T-4. T-4 was tested in 1968 in Novosibirsk, conducting the first quasistationary thermonuclear fusion reaction ever. The tokamak was dramatically more efficient than the other approaches of the same era, and most research after the 1970s concentrated on variations of this theme. The same is true today, where very large tokamaks like ITER are hoping to demonstrate several milestones on the way to commercial power production, including a burning plasma with long burn times, high power output and online fueling. There are no guarantees that the project will be successful, as previous generations of machines have faced formerly unseen problems on many occasions.

But the entire field of high temperature plasmas is much better understood now due to the earlier research, and there is considerable optimism that ITER will meet its goals. If successful, ITER would be followed by a "commercial demonstrator" system, similar to the very earliest power-producing fission reactors built in the era before wide-scale commercial deployment of larger machines started in the 1960s and 1970s. Even with these goals met, there are a number of major engineering problems remaining, notably finding suitable "low activity" materials for reactor construction, demonstrating secondary systems including practical tritium extraction, and building reactor designs that allow their reactor core to be removed when it becomes embrittled due to the neutron flux. Practical generators based on the tokamak concept remain far in the future.

The public at large has been somewhat disappointed, as the initial outlook for practical fusion power plants was much rosier than is now realized; a pamphlet from the 1970s printed by General Atomic stated that "Several commercial fusion reactors are expected to be online by the year 2000."

PINCH DEVICES

The Z-pinch phenomenon has been known since the end of the 18th century. Its use in the fusion field comes from research made on toroidal devices, initially in the Los Alamos National Laboratory right from 1952 (Perhapsatron), and in the United Kingdom from 1954 (ZETA), but its physical principles remained for a long time poorly understood and controlled. Pinch devices were studied as potential development paths to practical fusion devices through the 1950s, but studies of the data generated by these devices suggested that instabilities in the collapse mechanism would doom any pinch-type device to power levels that were far too low to suggest continuing along these lines would be practical.

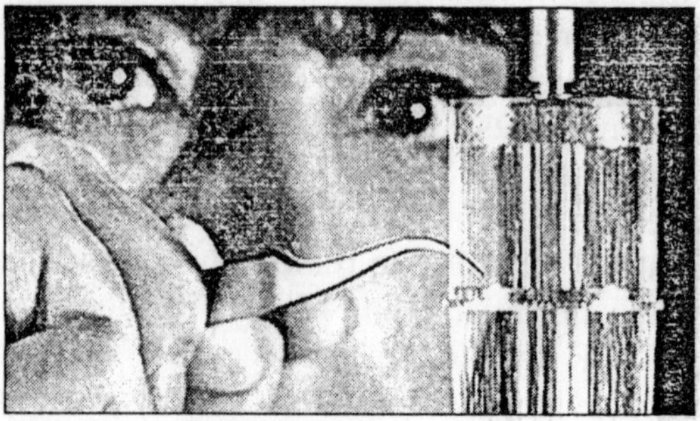

Fig. A "Wires Array" Used in Z-pinch Confinement, During the Building Process.

Most work on pinch-type devices ended by the 1960s. Recent work on the basic concept started as a result of the appearance of the "wires array" concept in the 1980s, which allowed a more efficient use of this technique. The Sandia National Laboratory runs a continuing wire-array research programme with the Zpinch machine.

In addition, the University of Washington's ZaP Lab have shown quiescent periods of stability hundreds of times longer than expected for plasma in a Z-pinch configuration, giving promise to the confinement technique.

Laser Inertial Devices

The technique of implosion of a microcapsule irradiated by laser beams, the basis of laser inertial confinement, was first suggested in 1962 by scientists at Lawrence Livermore National Laboratory, shortly after the invention of the laser itself in 1960. Lasers of the era were very low powered, but low-level research using them nevertheless started as early as 1965. More serious research started in the early 1970s when new types of lasers offered a path to dramatically higher power levels, levels that made inertial-confinement fusion devices appear practical for the first time. By the late 1970s great strides had been made in laser power, but with each increase new problems were found in the implosion technique that suggested even more power would be required. By the 1980s these increases were so large that using the concept for generating net energy seemed remote.

Most research in this field turned to weapons research, always a second line of research, as the implosion concept is somewhat similar to hydrogen bomb operation. Work on very large versions continued as a result, with the very large National Ignition Facility in the US and Laser Mégajoule in France supporting these research programs.

More recent work had demonstrated that significant savings in the required laser energy are possible using a technique known as "fast ignition". The savings are so dramatic that the concept appears to be a useful technique for energy production again, so much so that it is a serious contender for pre-commercial development. There are proposals to build an experimental facility dedicated to the fast ignition approach, known as HiPER. At the same time, advances in solid state lasers appear to improve the "driver" systems' efficiency by about ten times (to 10- 20 per cent), savings that make even the large "traditional" machines almost practical, and might make the fast ignition concept outpace the magnetic approaches in further development.

The laser-based concept has other advantages as well. The reactor core is mostly exposed, as opposed to being wrapped in a huge magnet as in the tokamak. This makes the problem of removing energy from the system somewhat simpler, and should mean that a laser-based device would be much easier to perform maintenance on, such as core replacement.

Additionally, the lack of strong magnetic fields allows for a wider variety of low-activation materials, including carbon fibre, which would reduce both the frequency of such neutron activations and the rate of irradiation the core. In other ways the programme has many of the same problems as the tokamak; practical methods of energy removal and tritium recycling need to be demonstrated, and in addition there is always the possibility that a new previously unseen collapse problem will arise.

Other Systems

Throughout the history of fusion power research there have been a number of devices that have produced fusion at a much smaller level, not being suitable for energy production, but nevertheless starting to fill other roles. Inventor of the Cathode Ray Tube Television Philo T. Farnsworth patented his first Fusor design in 1968, a device which uses inertial electrostatic confinement.

Towards the end of the 1960s, Robert Hirsch designed a variant of the Farnsworth Fusor known as the Hirsch-Meeks fusor. This variant is a considerable improvement over the Farnsworth design, and is able to generate neutron flux in the order of one billion neutrons per second. Although the efficiency was very low at first, there were hopes the device could be scaled up, but continued development demonstrated that this approach would be impractical for large machines. Nevertheless, fusion could be achieved using a "lab bench top" type set up for the first time, at minimal cost.

This type of fusor found its first application as a portable neutron generator in the late 1990s. An automated sealed reaction chamber version of this device, commercially named Fusionstar was developed by EADS but abandoned in 2001. Its successor is the NSD-Fusion neutron generator.

Robert W. Bussard's Polywell concept is roughly similar to the Fusor design, but replaces the problematic grid with a magnetically contained electron cloud which holds the ions in position and gives an accelerating potential. Bussard claimed that a scaled up version would be capable of generating net power. In April 2005, a team from UCLA announced it had devised a novel way of producing fusion using a machine that "fits on a lab bench", using lithium tantalate to generate enough voltage to smash deuterium atoms together. However, the process does not generate net power.

SAFETY AND THE ENVIRONMENT

The likelihood of a catastrophic accident in a fusion reactor in which injury or loss of life occurs is much smaller than that in a fission reactor. The primary reason is that the fission products in a fission reactor continue to generate heat through beta-decay for several hours or even days after reactor shutdown, meaning that a meltdown is possible even after the reactor has been stopped.

In contrast, fusion requires precisely controlled conditions of temperature, pressure and magnetic field parameters in order to generate net energy. If the reactor were damaged, these parameters would be disrupted and the heat generation in the reactor would rapidly cease.

There is also no risk of a runaway reaction in a fusion reactor, since the plasma is normally burnt at optimal conditions, and any significant change will render it unable to produce excess heat. In fusion reactors the reaction process is so delicate that this level of safety is inherent; no elaborate failsafe mechanism is required.

Although the plasma in a fusion power plant will have a volume of 1000 cubic meters or more, the density of the plasma is extremely low, and the total amount of fusion fuel in the vessel is very small, typically a few grams. If the fuel supply is closed, the reaction stops within seconds. In comparison, a fission reactor is typically loaded with enough fuel for one or several years, and no additional fuel is necessary to keep the reaction going.

In the magnetic approach, strong fields are developed in coils that are held in place mechanically by the reactor structure. Failure of this structure could release this tension and allow the magnet to "explode" outward. The severity of this event would be similar to any other industrial accident, and could be effectively stopped with a containment building similar to those used in existing (fission) nuclear generators.

The laser-driven inertial approach is generally lower-stress. Although failure of the reaction chamber is possible, simply stopping fuel delivery would prevent any sort of catastrophic failure.

Most reactor designs rely on the use of liquid lithium as both a coolant and a method for converting stray neutrons from the reaction into tritium, which is fed back into the reactor as fuel. Lithium is highly flammable, and in the case of a fire it is possible that the lithium stored on-site could be burned up and escape. In this case the tritium contents of the lithium would be released into the atmosphere, posing a radiation risk. However, calculations suggest that the total amount of tritium and other radioactive gases in a typical power plant would be so small, about 1 kg, that they would have diluted to legally acceptable limits by the time they blew as far as the plant's perimeter fence.

EFFLUENTS DURING NORMAL OPERATION

The natural product of the fusion reaction is a small amount of helium, which is completely harmless to life and does not contribute to global warming. Of more concern is tritium, which, like other isotopes of hydrogen, is difficult to retain completely. During normal operation, some amount of tritium will be continually released.

There would be no acute danger, but the cumulative effect on the world's population from a fusion economy could be a matter of concern. The 12 year half-life of tritium would at least prevent unlimited build-up and long-term contamination without appropriate containment techniques. Current ITER designs are investigating total containment facilities for any tritium.

Waste Management

The large flux of high-energy neutrons in a reactor will make the structural materials radioactive. The radioactive inventory at shut-down may be comparable to that of a fission reactor, but there are important differences. The half-life of the radioisotopes produced by fusion tend to be less than those from fission, so that the inventory decreases more rapidly. Furthermore, there are fewer unique species, and they tend to be non-volatile and biologically less active.

Unlike fission reactors, whose waste remains radioactive for thousands of years, most of the radioactive material in a fusion reactor would be the reactor core itself, which would be dangerous for about 50 years, and low-level waste another 100. Although this waste will be considerably more radioactive during those 50 years than fission waste, the very short half-life makes the process very attractive, as the waste management is fairly straightforward. By 300 years the material would have the same radioactivity as coal ash.

Additionally, the materials used in a fusion reactor are more "flexible" than in a fission design, where many materials are required for their specific neutron cross-sections. This allows a fusion reactor to be designed using materials that are selected specifically to be "low activation", materials that do not easily become radioactive.

Vanadium, for example, would become much less radioactive than stainless steel. Carbon fibre materials are also low-activation, as well as being strong and light, and are a promising area of study for laser-inertial reactors where a magnetic field is not required.

In general terms, fusion reactors would create far less radioactive material than a fission reactor, the material it would create is less damaging biologically, and the radioactivity "burns off" within a time period that is well within existing engineering capabilities.

Nuclear Proliferation

Although fusion power uses nuclear technology, the overlap with nuclear weapons technology is small. Tritium is a component of the trigger of hydrogen bombs, but not a major problem in production. The copious neutrons from a fusion reactor could be used to breed plutonium for an atomic bomb, but not without extensive redesign of the reactor, so that production would be difficult to conceal.

The theoretical and computational tools needed for hydrogen bomb design are closely related to those needed for inertial confinement fusion, but have very little in common with the more scientifically developed magnetic confinement fusion.

As a Sustainable Energy Source

Large-scale reactors using neutronic fuels (e.g. ITER) and thermal power production (turbine based) are most comparable to fission power from an engineering and economics viewpoint. Both fission and fusion power plants involve a relatively compact heat source powering a conventional steam turbine-based power plant, while producing enough neutron radiation to make activation of the plant materials problematic.

The main distinction is that fusion power produces no high-level radioactive waste (though activated plant materials still need to be disposed of). There are some power plant ideas which may significantly lower the cost or size of such plants; however, research in these areas is nowhere near as advanced as in tokamaks.

Fusion power commonly proposes the use of deuterium, an isotope of hydrogen, as fuel and in many current designs also use lithium. Assuming a fusion energy output equal to the 1995 global output of about 100 EJ/yr (= 1×10^{20} J/yr) and that this does not increase in the future, then the known current lithium reserves would last 3000 years, lithium from sea water would last 60 million years, and a more complicated fusion process using only deuterium from sea water would have fuel for 150 billion years.

THEORETICAL POWER PLANT DESIGNS

Confinement refers to all the conditions necessary to keep a plasma dense and hot long enough to undergo fusion:

- *Equilibrium:* There must be no net forces on any part of the plasma, otherwise it will rapidly disassemble. The exception, of course, is inertial confinement, where the relevant physics must occur faster than the disassembly time.
- *Stability:* The plasma must be so constructed that small deviations are restored to the initial state, otherwise some unavoidable disturbance will occur and grow exponentially until the plasma is destroyed.
- *Transport:* The loss of particles and heat in all channels must be sufficiently slow. The word "confinement" is often used in the restricted sense of "energy confinement".

The first human-made, large-scale production of fusion reactions was the test of the hydrogen bomb, Ivy Mike, in 1952. As part of the PACER (fusion) project, it was once proposed to use hydrogen bombs as a source of power by detonating them in underground caverns and then generating electricity from the heat produced, but such a power plant is unlikely ever to be constructed, for a variety of reasons.

Controlled thermonuclear fusion (CTF) refers to the alternative of continuous power production, or at least the use of explosions that are so small that they do not destroy a significant portion of the machine that produces them. To produce self-sustaining fusion, the energy released by the reaction (or at least a fraction of it) must be used to heat new reactant nuclei and keep them hot long enough that they also undergo fusion reactions.

Retaining the heat is called energy confinement and may be accomplished in a number of ways. The hydrogen bomb really has no confinement at all. The fuel is simply allowed to fly apart, but it takes a certain length of time to do this, and during this time fusion can occur. This approach is called inertial confinement.

If more than milligram quantities of fuel are used (and efficiently fused), the explosion would destroy the machine, so theoretically, controlled thermonuclear fusion using inertial confinement would be done using tiny pellets of fuel which explode several times a second. To induce the explosion, the pellet must be compressed to about 30 times solid density with energetic beams.

If the beams are focused directly on the pellet, it is called direct drive, which can in principle be very efficient, but in practice it is difficult to obtain the needed uniformity. An alternative approach is indirect drive, in which the beams heat a shell, and the shell radiates x-rays, which then implode the pellet.

The beams are commonly laser beams, but heavy and light ion beams and electron beams have all been investigated. Inertial confinement produces plasmas with impressively high densities and temperatures, and appears to be best suited to weapons research, X-ray generation, very small reactors, and perhaps in the distant future, spaceflight. They rely on fuel pellets with close to a "perfect" shape in order to generate a symmetrical inward shock wave to produce the high-density plasma, and in practice these have proven difficult to produce.

A recent development in the field of laser induced ICF is the use of ultrashort pulse multi-petawatt lasers to heat the plasma of an imploding pellet at exactly the moment of greatest density after it is imploded conventionally using terawatt scale lasers.

At the temperatures required for fusion, the fuel is in the form of a plasma with very good electrical conductivity. This opens the possibility to confine the fuel and the energy with magnetic fields, an idea known as magnetic confinement. The Lorenz force works only perpendicular to the magnetic field, so that the first problem is how to prevent the plasma from leaking out the ends of the field lines. There are basically two solutions.

The first is to use the magnetic mirror effect. If particles following a field line encounter a region of higher field strength, then some of the particles will be stopped and reflected. Advantages of a magnetic mirror power plant would be simplified construction and maintenance due to a linear topology and the potential to apply direct conversion in a natural way, but the confinement achieved in the experiments was so poor that this approach has been essentially abandoned.

The second possibility to prevent end losses is to bend the field lines back on themselves, either in circles or more commonly in nested toroidal surfaces. The most highly developed system of this type is the *tokamak,* with the *stellarator* being next most advanced, followed by the Reversed field pinch. Compact toroids, especially the *Field-Reversed Configuration* and the spheromak, attempt to combine the advantages of toroidal magnetic surfaces with those of a simply connected (non-toroidal) machine, resulting in a mechanically simpler and smaller confinement area.

Compact toroids still have some enthusiastic supporters but are not backed as readily by the majority of the fusion community. Finally, there are also *electrostatic confinement fusion* systems, in which ions in the reaction chamber are confined and held at the centre of the device by electrostatic forces, as in the Farnsworth-Hirsch Fusor, which is not believed to be able to be developed into a power plant.

The Polywell, an advanced variant of the fusor, has shown a degree of research interest as of late; however, the technology is relatively immature, and major scientific and engineering questions remain which researchers under the auspices of the U.S. Office of Naval Research hope to further investigate.

Other Approaches

A more subtle technique is to use more unusual particles to catalyse fusion. The best known of these is Muon-catalyzed fusion which uses muons, which behave somewhat like electrons and replace the electrons around the atoms. These muons allow atoms to get much closer and thus reduce the kinetic energy required to initiate fusion.

Muons require more energy to produce than we can get back from muon-catalysed fusion, making this approach impractical for the generation of power. Some researchers have reported excess heat, neutrons, tritium, helium and other nuclear effects in so-called cold fusion systems. In 2004, a peer review panel was commissioned by the US Department of Energy to study these claims.

This identified basic areas of research which were necessary for acceptance of the idea, but did not recommend a Federally-funded programme. Research into sonoluminescence induced fusion, sometimes known as *"bubble fusion"*, also continues, although it is met with as much skepticism as cold fusion is by most of the scientific community.

Subsystems

In fusion research, achieving a fusion energy gain factor $Q = 1$ is called breakeven and is considered a significant although somewhat artificial milestone. Ignition refers to an infinite Q, that is, a self-sustaining plasma where the losses are made up for by fusion power without any external input.

In a practical fusion reactor, some external power will always be required for things like current drive, refueling, profile control, and burn control.

A value on the order of $Q = 20$ will be required if the plant is to deliver much more energy than it uses internally. There have been many design studies for fusion power plants. Despite many differences, there are several systems that are common to most. To begin with, a fusion power plant, like a fission power plant, is customarily divided into the nuclear island and the balance of plant. The balance of plant is the conventional part that converts high-temperature heat into electricity via steam turbines.

It is much the same in a fusion power plant as in a fission or coal power plant. In a fusion power plant, the nuclear island has a plasma chamber with an associated vacuum system, surrounded by plasma-facing components (first wall and divertor) maintaining the vacuum boundary and absorbing the thermal radiation coming from the plasma, surrounded in turn by a blanket where the neutrons are absorbed to breed tritium and heat a working fluid that transfers the power to the balance of plant.

If magnetic confinement is used, a magnet system, using primarily cryogenic superconducting magnets, is needed, and usually systems for heating and refueling the plasma and for driving current. In inertial confinement, a driver (laser or accelerator) and a focusing system are needed, as well as a means for forming and positioning the pellets.

Although the standard solution for electricity production in fusion power plant designs is conventional steam turbines using the heat deposited by neutrons, there are also designs for direct conversion of the energy of the charged particles into electricity. These are of little value with a D-T fuel cycle, where 80 per cent of the power is in the neutrons, but are indispensable with aneutronic fusion, where less than 1 per cent is.

Direct conversion has been most commonly proposed for open-ended magnetic configurations like magnetic mirrors or Field-Reversed Configurations, where charged particles are lost along the magnetic field lines, which are then expanded to convert a large fraction of the random energy of the fusion products into directed motion. The particles are then collected on electrodes at various large electrical potentials. Typically the claimed conversion efficiency is in the range of 80 per cent, but the converter may approach the reactor itself in size and expense.

Fusion Materials Irradiation

Developing materials for fusion reactors has long been recognized as a problem nearly as difficult and important as that of plasma confinement, but it has received only a fraction of the attention. The neutron flux in a fusion reactor is expected to be about 100 times that in existing pressurized water reactors (PWR). Each atom in the blanket of a fusion reactor is expected to be hit by a neutron and displaced about a hundred times before the material is replaced.

Furthermore the high-energy neutrons will produce hydrogen and helium in various nuclear reactions that tends to form bubbles at grain boundaries and result in swelling, blistering or embrittlement. One also wishes to choose materials whose primary components and impurities do not result in long-lived radioactive wastes. Finally, the mechanical forces and temperatures are large, and there may be frequent cycling of both.

The problem is exacerbated because realistic material tests must expose samples to neutron fluxes of a similar level for a similar length of time as those expected in a fusion power plant. Such a neutron source is nearly as complicated and expensive as a fusion reactor itself would be. Proper materials testing will not be possible in ITER, and a proposed materials testing facility, IFMIF, was still at the design stage in 2005.

The material of the plasma facing components (PFC) is a special problem. The PFC do not have to withstand large mechanical loads, so neutron damage is much less of an issue. They do have to withstand extremely large thermal loads, up to 10 MW/m^2, which is a difficult but solvable problem. Regardless of the material chosen, the heat flux can only be accommodated without melting if the distance from the front surface to the coolant is not more than a centimeter or two. The primary issue is the interaction with the plasma.

One can choose either a low-Z material, typified by graphite although for some purposes beryllium might be chosen, or a high-Z material, usually tungsten with molybdenum as a second choice. Use of liquid metals (lithium, gallium, tin) has also been proposed, e.g., by injection of 1-5 mm thick streams flowing at 10 m/s on solid substrates.

If graphite is used, the gross erosion rates due to physical and chemical sputtering would be many meters per year, so one must rely on redeposition of the sputtered material. The location of the redeposition will not exactly coincide with the location of the sputtering, so one is still left with erosion rates that may be prohibitive.

An even larger problem is the tritium co-deposited with the redeposited graphite. The tritium inventory in graphite layers and dust in a reactor could quickly build up to many kilograms, representing a waste of resources and a serious radiological hazard in case of an accident.

The consensus of the fusion community seems to be that graphite, although a very attractive material for fusion experiments, cannot be the primary PFC material in a commercial reactor. The sputtering rate of tungsten can be orders of magnitude smaller than that of carbon, and tritium is not so easily incorporated into redeposited tungsten, making this a more attractive choice.

On the other hand, tungsten impurities in a plasma are much more damaging than carbon impurities, and self-sputtering of tungsten can be high, so it will be necessary to ensure that the plasma in contact with the tungsten is not too hot (a few tens of eV rather than hundreds of eV). Tungsten also has disadvantages in terms of eddy currents and melting in off-normal events, as well as some radiological issues.

Chapter 15

Systems of Nuclear Instrumentation

In a car we need to know how fast we are driving so we do not exceed the speed limit. In a nuclear power plant, the operator must know how much neutron power is being produced so that he or she does not exceed the regulatory power limit.

Neutron Power level (or power level, for short) can actually span about 11 orders of magnitude (11 powers of 10) from the lowest power level to full power. No one type of detector can be used. In many plants, 3 levels, and methods, of detection are used. These levels are called:

- Source Range (usually a proportional detector)
- Intermediate Range (usually a compensated ion chamber)
- Power Range (usually an uncompensated ion chamber)

The neutron detectors, used by the operator, may be installed inside or outside the reactor, depending on the type of reactor. If detectors normally used by the operator are outside the reactor, these detectors must be periodically recalibrated with detectors that are either installed in the reactor core or that pass through the reactor core.

The neutron detectors are typically cylindrical with a central wire. The wire and the outside of the detector have voltage differences between 10 and 2500 volts dc. The detector often has a gas that will interact with the neutron to produce a charged particle. The charged particles then are drawn to the appropriate central wire or outside of the detector. These events produce either a single pulse or a current depending on the type of detector. Gases used in neutron detectors usually include boron or uranium.

When a neutron reacts with boron, it produces 2 charged particles - an alpha particle (helium nucleus $_2He^4$ with a +2 charge). When a neutron reacts with the uranium it fissions creating 2 fission products that are both highly positively charged.

TRENDS IN NUCLEAR INSTRUMENTATION IN DIAGNOSTIC NUCLEAR MEDICINE

In nuclear medicine imaging the use of multi-detector systems for total-body, brain and heart scanning have recently gained increasing popularity. The systems with 2 or 3 Anger type gamma cameras 180°, 120° or 90° apart with small or large filed of view are most popular.

These systems have considerably improved the sensitivity, resolution and the scanning time comparing to the single camera systems. PET imaging with rings of small detectors in multi-slice configuration with lead or tungsten septa between slices (2-D) or without septa (3-D) has been successfully implemented using FDG in a limited axial region and most recently also for total-body tomography.

The biggest improvement in spatial resolution of PET in the past was gained mainly by the reducing the size of the crystals which currently resolve the structures to 5 mm in size. There is a strong development of the imaging systems with a large number of tiny crystals made by the newly developed high dense and fast responding materials.

Principles of Image Detection

Generally three basic methods of nuclear medicine image formation are actual: 1. single large scintillation crystal with a large number of photo multiplier tubes (PMT), 2. large number of tiny scintillation crystals with position sensitive photo-multiplier tube (PSPMT) and 3. large semiconductor crystal with array of tiny "n-p" sensitive areas.

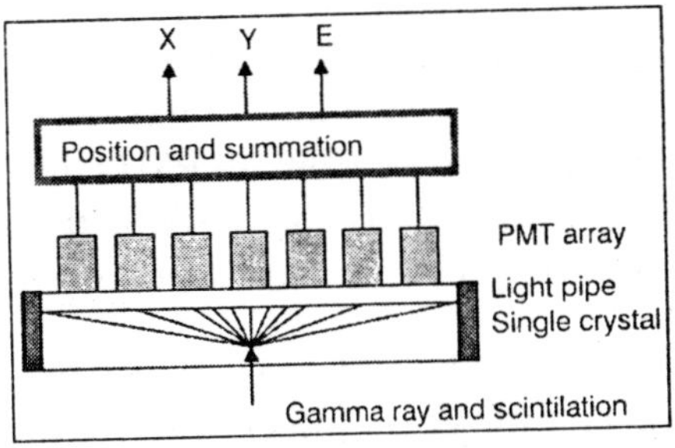

Fig. Imaging Detector with Single Large Scintillation Crystal with Set of PMTs.

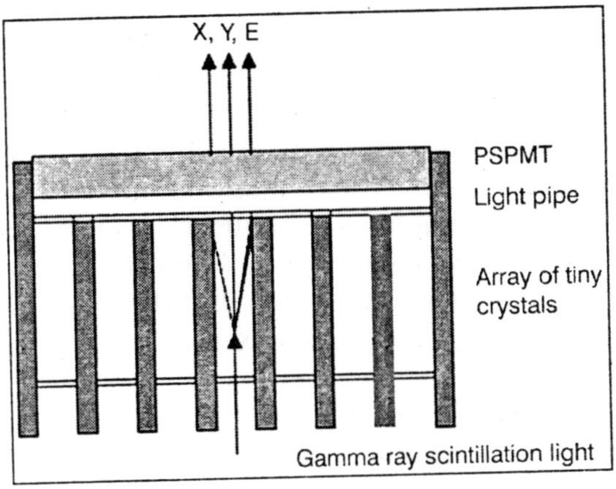

Fig. Imaging Detector with Array of Tiny Scintillation Crystals with PSPMT.

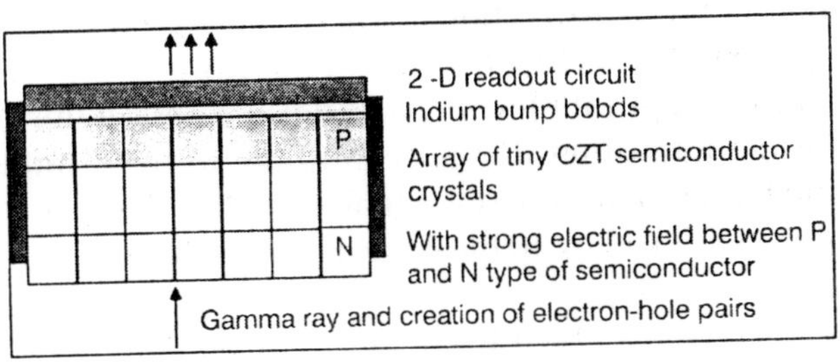

Fig. Imaging Semiconductor Array Detector.

In the first method the gamma ray strikes a large circular or rectangle thin crystal and the induced scintillation light is distributed between the PMTs according to their viewing spatial angles. The x and y positions of PMTs are weighted by their electric signal responses from all PMTs and the X and Y coordinates of the scintillation cloud striking the array of PMT and the corresponding energy is computed (Anger type of gamma camera).

The intrinsic spatial resolution of the imaging device strongly depends on the crystal thickness, slightly less on the size / number / shape of PMT and the position circuit. The sensitivity of the system oppositely to the spatial resolution increases by the crystal thickness. The energy and time resolution depend mainly on the crystal material. Nowadays nearly all planar gamma cameras are of this kind.

In the second method the gamma ray is absorbed in a tiny crystal and all the induced scintillation light is collected by a small area of position sensitive photo multiplier tube which converts the incident light in a very thin layer into a charge or current which is then converted to digital E (energy) signal. Each small sensitive area of PSPMT provides also corresponding spatial coordinates X and Y for the particular exposed crystal.

The spatial resolution strongly depends on the size of the crystals and on the thickness and material of the septa. Each crystal represents a pixel in the final digital image. Thinner and longer the crystal and thinner the septa better is the resolving power of the imaging system. Still, there is a limitation in the cross sectional size of the crystal.

If gamma ray strikes the crystal too close to the reflector cover or to the optical fibre (crystal is inside optical fibre) then some of the ionization doesn't contribute to the scintillation light and therefore the energy signal is smaller. Because of this, a definite volume close to the edge, is not useful and is treated as scattered.

An estimate is given for the circular shape of the crystal and for two popular scintillation materials (NaI and BGO). For approximately 20 per cent of absorbed gamma rays, the reduction of the scintillation light will be present. Some of the signals coming from these 20 per cent will still be included in the lower part of the photopeak but some will be lost. The conclusion is that there is no meaning of using thinner sized crystals than 1-2 mm depending on the material (for NaI crystal this size is limited to 2 mm and for BGO crystal 1 mm). The sensitivity increases by the crystal length, density and cross sectional size. The energy resolution is improved by more efficient PSPMT and more effective collecting of scintillation light in crystal.

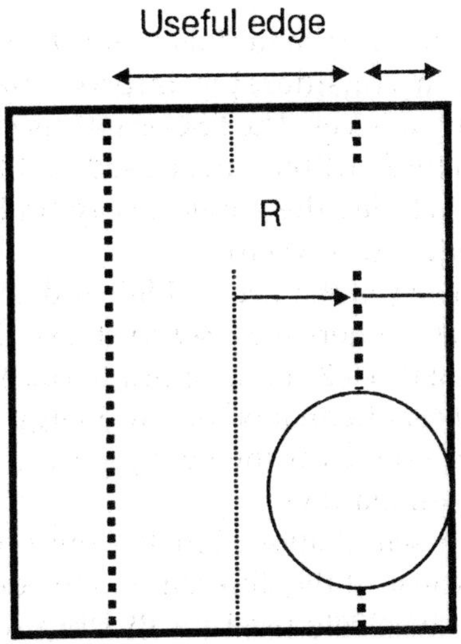

Fig. Estimate of the limit for the crystal cross sectional size.

In the third method the incident gamma ray is absorbed in the region of "p-n junction" region of the semiconductor crystal and a large number of electron-hole pairs are created. Their number (approximately 3 - 5 eV/electron-hole pair is spent on average) is proportional to the energy of the gamma ray and is nearly ten times greater than the quantity of the scintillation light (approximately 30 eV per ionization).

For the same factor the energy resolution is better. The efficiency of the late developed CZT crystal is even better than for NaI. The intrinsic spatial resolution of the CZT gamma camera is considerably better and is about 2 - 3 mm for 140 keV compared to 3 - 4 mm for NaI gamma camera. On the other hand the collection time for electron-hole pairs is at least

100 times shorter than is the decay time for scintillation light in NaI crystal and therefore the increased count rate can be achieved (250.000 counts/s). The weight of such imaging system is 100 times lower and the CZT gamma camera is easily portable to emergency departments or elsewhere. It is expected that the price for such gamma camera will also be much lower because of the less complicated production.

New Materials for Detection Crystals

Some new detector materials were developed recently which promise a considerable improvement of nuclear medicine imaging devices. The LSO is intrinsically radioactive and is not useful for SPET but can be used for PET (coincidence measurement excludes the single decay and absorption of gamma ray inside LSO crystal).

Very promising is the use of LSO and YSO in so-called phoswhich detector where the YSO crystal (1 - 2 cm) is in front and the LSO crystal (1 - 2 cm) is optically coupled to the YSO. The YSO is used for attenuation of low energy (in the range of 100 keV) and LSO serves as the light pipe and for attenuation of high energy gamma rays.

The induced scintillation signals from both crystals can be separated because of the difference in the decay times. The BGO is nearly exclusively used for PET but will probably be replaced by this phoswhich detector. One of the most interesting detectors becomes semiconductor CZT (cadmium zinc telluride) which has even better stopping power for 140 keV gamma rays than NaI (Tl) but much better energy resolution (for factor of 10). The array with large number of very small sensitive areas can be build so that each of these areas is a pixel in the digital image.

SPET

The biggest advance in the SPET was the implementation of several detector heads which drastically improved the system sensitivity. For cardiac SPET the use of two heads at 90° and the whole-body bone SPET or scanning at 180° shortens the acquisition time or increases the acquisition counts two times. The improvement of the gamma camera features was mainly due to the development of so-called digital head electronics which had replaced the old analog position circuit by the digital one.

The output from each PMT is digitized and then the spatial coordinates are computed. All corrections for non-linearity, spatial and energy non-uniformity can be performed on-line by the use of special fast processors. In the future the probable development of SPET will be building the tomographic system from the arrangement of several modular multi-crystal detectors which will introduce even greater flexibility than with two or three big planar gamma cameras at different angular setup.

Each module will probably be an array of very tiny crystals from YSO and LSO or semiconductor CZT detector. Another possible approach in modular design of SPET will be in a specialization for some organs (i.e. thyroid and cardiac tomograph needs relatively small sized detector's modules) and possible much lower prices for small SPET systems.

PET

The latest improvements are mainly due to the development of the so called 3-D tomographs which don't use septa between planes with rings of crystals. By omitting the septa the sensitivity is increased by ten times and the amount of scattered photons only approximately 3o per cent. In such configuration of several thousands tiny crystals all possible coincidence events between any two crystals are used in the reconstruction algorithm. The system works in a true 3-D mode.

Another possibility of PET is the use of dual-head SPET system with or without collimator (high-speed electronics is essential) in a coincidence mode. But this type of PET is considerably less sensitive but is interesting for performing both PET and SPET studies. A much better sensitivity of the system is expected in future from the new generation of the PET. It will be of extreme importance in the imaging of specific biochemical bindings, such as receptor binding. In this applications a small amount of the injected radioactivity is collected by a target organ (usually less than 1 per cent).

Surgical Gamma Probe

This application of radioactivity tracing becomes very important in surgery for identifying the regional metastases. Currently interesting clinical field where the small detector probe is of great importance is lymph node dissection of the axilla or regional nodes in the breast cancer and in some melanoma patients.

The role of the surgical gamma probe is to localise the sentinel node transcutaneously and intra-operatively. To meet a high sensitivity, good spatial and spectral resolution and appropriate ergonomic characteristics several different commercially available probes were evaluated (4). It was found that CZT probe was the most appropriate for low energies (140 keV from 99mTc) and the NaI probe for high energy (364 keV 131I).

Nuclear Medicine instrumentation is passing vigorous development last years and most likely in the near future as well. New detection materials with much better physical characteristics than the standard NaI concerning the stopping power, energy resolution, fragility, decay time, light output, and density will most likely replace the NaI.

It is expected that new imaging devices with several thousands of tiny crystals or semiconductor array of small position sensitive areas will improve the sensitivity and specificity of clinical studies. At the same time the small surgical probes made from these materials are also becoming very popular in surgery tracing the regional metastases.

MAGNETIC RESONANCE IMAGING

Medical Imaging

Medical imaging refers to the techniques and processes used to create images of the human body (or parts thereof) for clinical purposes (medical procedures seeking to reveal, diagnose or examine disease) or medical science (including the study of normal anatomy and function). As a discipline and in its widest sense, it is part of biological imaging and incorporates radiology (in the wider sense), radiological sciences, endoscopy, (medical) thermography, medical photography and microscopy (e.g. for human pathological investigations).

Measurement and recording techniques which are not primarily designed to produce images, such as electroencephalography (EEG) and magnetoencephalography (MEG) and others, but which produce data susceptible to be represented as maps (i.e. containing positional information), can be seen as forms of medical imaging.

In the clinical context, medical imaging is generally equated to Radiology or "clinical imaging" and the medical practitioner responsible for interpreting (and sometimes acquiring) the images is a radiologist. Diagnostic radiography designates the technical aspects of medical imaging and in particular the acquisition of medical images. The *radiographer* or *radiologic technologist* is usually responsible for acquiring medical images of diagnostic quality, although some radiological interventions are performed by radiologists.

As a field of scientific investigation, medical imaging constitutes a sub-discipline of biomedical engineering, medical physics or medicine depending on the context: Research and development in the area of instrumentation, image acquisition (e.g. radiography), modelling and quantification are usually the preserve of biomedical engineering, medical physics and computer science; Research into the application and interpretation of medical images is usually the preserve of radiology and the medical sub-discipline relevant to medical condition or area of medical science (neuroscience, cardiology, psychiatry, psychology, etc) under investigation.

Many of the techniques developed for medical imaging also have scientific and industrial applications. Medical imaging is often perceived to designate the set of techniques that noninvasively produce images of the internal aspect of the body. In this restricted sense, medical imaging can be seen as the solution of mathematical inverse problems. This means that cause (the properties of living tissue) is inferred from effect (the observed signal). In the case of ultrasonography the probe consists of ultrasonic pressure waves and echoes inside the tissue show the internal structure. In the case of projection radiography, the probe is X-ray radiation which is absorbed at different rates in different tissue types such as bone, muscle and fat.

MODERN IMAGING TECHNOLOGY

Fluoroscopy

Fluoroscopy produces real-time images of internal structures of the body in a similar fashion to radiography, but employs a constant input of x rays. Contrast media, such as barium, iodine, and air are used to visualize internal organs as they work. Fluoroscopy is also used in image-guided procedures when constant feedback during a procedure is required.

Magnetic Resonance Imaging (MRI)

A Magnetic Resonance Imaging instrument (MRI scanner) uses powerful magnets to polarise and excite hydrogen nuclei (single proton) in water molecules in human tissue, producing a detectable signal which is spatially encoded resulting in images of the body.

In brief, MRI involves the use of three kinds of electromagnetic field: a very strong (of the order of units of teslas) static magnetic field to polarize the hydrogen nuclei, called the static field; a weaker time-varying (of the order of 1 kHz) for spatial encoding, called the gradient field(s); and a weak radio-frequency (RF) field for manipulation of the hydrogen nuclei to produce measurable signals, collected through an RF antenna.

Like CT, MRI traditionally creates a 2D image of a thin "slice" of the body and is therefore considered a tomographic imaging technique. Modern MRI instruments are capable of producing images in the form of 3D blocks, which may be considered a generalisation of the single-slice, tomographic, concept. Unlike CT, MRI does not involve the use of ionizing radiation and is therefore not associated with the same health hazards; for example there are no known long term effects of exposure to strong static fields and therefore there is no limit on the number of scans to which an individual can be subjected, in contrast with X-ray and CT.

However, there are well identified health risks associated with tissue heating from exposure to the RF field and the presence of implanted devices in the body, such as pace makers. These risks are strictly controlled as part of the design of the instrument and the scanning protocols used. CT and MRI being sensitive to different properties of the tissue, the appearance of the images obtained with the two techniques differ markedly.

In CT, X-rays must be blocked by some form of dense tissue to create an image, therefore the image quality when looking at soft tissues will be poor. While any nucleus with a net nuclear spin can be used, the proton of the hydrogen atom remains the most widely used, especially in the clinical setting, since it is so ubiquitous and returns much signal. This nucleus, present in water molecules, allows excellent soft-tissue contrast.

MRI, or "NMR imaging" as it was originally known, has only been in use since the early 1980s. Effects from long term, or repeated exposure, to the intense static magnetic field are not known.

Nuclear Medicine

Images from gamma cameras are used in Nuclear Medicine to detect regions of biological activity that are often associated with diseases. A short lived isotope, such as ^{131}I is administered to the patient. These isotopes are more readily absorbed by biologically active regions of the body, such as tumors or fracture points in bones.

Positron Emission Tomography (PET)

Positron emission tomography is primarily used to detect diseases of the brain and heart. Similarly to nuclear medicine, a short-lived isotope, such as ^{18}F, is incorporated into a substance used by the body such as glucose which is absorbed by the tumor of interest.PET scans are often viewed along side computed tomography scans, which can be performed on the same equipment without moving the patient. This allows the tumors detected by the PET scan to be viewed next to the rest of the patient's anatomy detected by the CT scan.

Projection Radiography

Radiographs, more commonly known as x-rays, are often used to determine the type and extent of a fracture as well as for detecting pathological changes in the lungs. With the use of radio-opaque contrast media, such as barium, they can also be used to visualize the structure of the stomach and intestines - this can help diagnose ulcers or certain types of colon cancer.

Photoacoustic Imaging

Photoacoustic imaging is a recently developed hybrid biomedical imaging modality based on the photoacoustic effect. It combines the advantages of optical absorption contrast with ultrasonic spatial resolution for deep imaging in (optical) diffusive or quasi-diffusive regime. Recent studies have shown that photoacoustic imaging can be used in vivo for tumor angiogenesis monitoring, blood oxygenation mapping, functional brain imaging, and skin melanoma detection etc.

TOMOGRAPHY

Tomography is the method of imaging a single plane, or slice, of an object resulting in a tomogram. There are several forms of tomography:

Linear Tomography:

This is the most basic form of tomography. The X-ray tube moved from point "A" to point "B" above the patient, while the cassette holder (or "bucky") moves simultaneously under the patient from point "B" to point "A." The fulcrum, or pivot point, is set to the area of interest. In this manner, the points above and below the focal plane are blurred out, just as the background is blurred when panning a camera during exposure. No longer carried out and replaced by computed tomography.

Poly Tomography:

This was a complex form of tomography. With this technique, a number of geometrical movements were programmed, such as hypocycloidic, circular and elliptical. Philips Medical Systems produced one such device called the 'Polytome.' No longer carried out, replaced by computed tomography.

Zonography:

This is a variant of linear tomography, where a limited arc of movement is used. It is still used in some centres for visualising the kidney during an intravenous urogram (IVU)

Orthopantomography

The only common tomographic examination in use. This makes use of a complex movement to allow the radiographic examination of the mandible, as if it were a flat bone. It is often referred to as a "Panaray", but this is incorrect, as it is a trademark of a specific company's equipment

Computed Tomography (CT)

A CT scan, also known as a CAT scan, is a helical tomography (latest generation), which traditionally produces a 2D image of the structures in a thin section of the body. It uses X-rays. It has a greater ionizing radiation dose burden than projection radiography; repeated scans should be limited.

CLINICAL IMAGING OR BIOLOGICAL IMAGING TECHNIQUES

Electron Microscopy

The electron microscope is a microscope that can magnify very small details with high resolving power due to the use of electrons as the source of illumination, magnifying at levels up to 2,000,000 times. Electron microscopy is employed in anatomic pathology to identify organelles within the cells.

Its usefulness has been greatly reduced by immunhistochemistry but it is still irreplaceable for the diagnosis of kidney disease, identification of immotile cilia syndrome and many other tasks

Creation of Three-dimensional Images

Recently, techniques have been developed to enable CT, MRI and ultrasound scanning software to produce 3D images for the physician. Traditionally CT and MRI scans produced 2D static output on film. To produce 3D images, many scans are made, then combined by computers to produce a 3D model, which can then be manipulated by the physician. 3D ultrasounds are produced using a somewhat similar technique.

With the ability to visualize important structures in great detail, 3D visualization methods are a valuable resource for the diagnosis and surgical treatment of many pathologies.

It was a key resource for the famous, but ultimately unsuccessful attempt by Singaporean surgeons to separate Iranian twins Ladan and Laleh Bijani in 2003. The 3D equipment was used previously for similar operations with great success.

Other proposed or developed techniques include:

- Diffuse optical tomography
- Elastography
- Electrical impedance tomography
- Optoacoustic imaging
- Ophthalmology
  - A-scan
  - B-scan
  - Corneal topography
  - Heidelberg retinal tomography
  - Optical coherence tomography
  - scanning laser ophthalmoscopy

Some of these techniques are still at a research stage and not yet used in clinical routines.

NON-DIAGNOSTIC IMAGING

Neuroimaging has also been used in experimental circumstances to allow people (especially disabled persons) to control outside devices, acting as a brain computer interface.

Medical Imaging Service

This is a specialized area of medical equipment service and repair, which is separate from the biomedical field, although a hospital with their own service group may include them in the biomed department.

At one time, there were only two ways to receive training for this field. One was to learn it in the military, and the other was on-the-job training (OJT) from the manufacturer. But since the 1980s several independent training centers have been started. One such school is RSTI.

There are different means of employment in this occupation. Working for the manufacturer's field service department (OEM), working for a hospital (in-house), and working for an independent (outside, or independent provider). The most stable positions are with the OEM or hospital, as we can remain current through on-going training, and the two have good working relationships.

The OEM service engineer can expect to spend a lot of time driving from one site to another during the work day, and working non-standard hours. They will install, remove, diagnose, repair, calibrate, perform preventive maintenance, and interface equipment, all while ensuring good customer relations. We may also be required to do yearly testing of the radiation sources for Federal and State compliance.

The in-house person is employed by the hospital. With larger medical facilities, travel between the hospitals' other locations may be required to perform the required services. We may also be required to do yearly testing of the radiation sources for compliance.

The OEM or independent will provide installation of purchased equipment, and can be used for back-up service. An independent is typically someone who has left an OEM, and started their own service business. Staying up-to-date as an independent can be difficult and expensive, as the OEM is usually reluctant to provide training. However, non-OEM training facilities are available, such as the aforementioned RSTI. Competition for service can be aggressive, with OEM's giving hospitals or clinics a reduction in equipment purchase price if they retain some form of OEM service.

The independent may also sell and install refurbished equipment, or de-install equipment. They will repair, calibrate, and perform preventative maintenance. Because many of the tasks associated with imaging service require expensive, specialized equipment, there may be a financial limit to the independent.

Typical equipment used routinely are a Storage Oscilloscope and multimeter (if servicing old vacuum-state equipment, a VOM would be helpful). Additional equipment: Keithley dosimeter, mAs meter, Biddle contact tachometer, light to radiation template, etc.

Magnetic resonance imaging (MRI) is primarily used in medical imaging to visualize the structure and function of the body. It provides detailed images of the body in any plane. MR has much greater soft tissue contrast than Computed tomography (CT) making it especially useful in neurological, musculoskeletal, cardiovascular and oncolological diseases. Unlike CT it uses no ionizing radiation.

The scanner creates a powerful magnetic field which aligns the magnetization of hydrogen atoms in the body. Radio waves are used to alter the alignment of this magnetization. This causes the hydrogen atoms to emit a weak radio signal which is amplified by the scanner. This signal can be manipulated by additional magnetic fields to build up enough information to reconstruct an image of the body.

Magnetic resonance spectroscopy is used to measure the levels of different metabolites in body tissues. The MR signal produces spectrum of difference resonances that correspond to different molecular arrangements of the isotope being "excited". This signature is used to diagnose certain metabolic disorders, especially those affecting the brain, as well as to provide information on tumor metabolism.

The scanners used in medicine have a typical magnetic field strength of 0.2 to 3 teslas. Construction costs approximately US$ 1 million per tesla and maintenance an additional several hundred thousand dollars per year. Research using MRI scanners operating at ultra high field strength (up to 21.1 tesla) can produce images of the mouse brain with a resolution of 18 micrometres.

Nomenclature

Magnetic resonance imaging was developed from knowledge gained in the study of nuclear magnetic resonance. In its early years MRI was referred to as nuclear magnetic resonance imaging (NMRI), but the word nuclear has been associated with ionizing radiation exposure, which is not used in an MRI, so to prevent patients from making a negative association between MRI and ionizing radiation, the word has been almost universally removed.

Scientists still use the term *NMRI* when discussing non-medical devices operating on the same principles. One of the inventors of MRI, Paul Lauterbur, originally named the technique *zeugmatography*, a Greek term meaning "that which is used for joining". The term referred to the interaction between the static and the gradient magnetic fields necessary to create an image, but the nomenclature never caught on.

Brief Explanation for the Layperson

MRI works by making certain atoms in the body emit radiowaves by using a powerful magnet. When a person is in the scanner some of the hydrogen atoms, which are mainly found in water, align with this magnetic field. A radiowave at just the right frequency makes these atoms resonate and absorb energy. The atoms then release this energy in the form of a very weak radiowave that can be measured by the scanner and amplified.

Extra magnetic fields are used to constantly change the magnetic field to allow images of the body to be reconstructed. These fields are created by gradient coils which make the familiar banging sounds of an MRI scan. Contrast agents may be injected to demonstrate blood vessels or inflammation in the tissues. Unlike CT scanning MRI uses no ionizing radiation and is generally a very safe procedure. Patients with some metal implants and cardiac pacemakers are prevented from having an MRI due to effect of the powerful magnetic field.

MRI is used to image every part of the body, but is particularly useful in neurological conditions, disorders of the muscles and joints, for evaluating tumors and showing abnormalities in the heart and blood vessels.

PRINCIPLE

Magnetism

Elementary subatomic particles such as protons have the quantum mechanical property of spin.

Nuclei such as ^{1}H or ^{31}P, with an odd number of nucleons, always have a non–zero spin and therefore a magnetic moment. Some other isotopes such as ^{12}C have no unpaired neutrons or protons, and no net spin.

When these spins are placed in an external magnetic field they start to precess around the direction of that field. The magnetic field also creates two energy states that the protons can occupy which are separated by a quantum of energy. The thermal energy of the sample causes the molecules to tumble leaving only a very small excess of protons to cause magnetic polarization.

Resonance

The energy difference between the proton energy states corresponds to electromagnetic radiation at radio frequency wavelengths. Resonant absorption of energy by the protons due to an external oscillating magnetic field (radio wave) will occur at the Larmor frequency.

The net magnetization vector has two components. The longitudinal magnetization is due to an excess of protons in the lower energy state. This gives a net polarization parallel to the external field. The transverse magnetization is due to coherences forming between the two proton energy states. This gives a net polarization perpendicular to external field in the transverse plane. The recovery of longitudinal magnetization is called T1 relaxation and the loss of phase coherence in the transverse plane is called T2 relaxation.

When the radio frequency pulse is turned off, the transverse vector component produces an oscillating magnetic field which induces a small current in the receiver coil. This free induction decay (FID) lasts only a few milliseconds before the thermal equilibrium of the spins is restored. The actual signal that is measured by the scanner is formed by a refocusing gradient or radio wave to create a gradient or spin-echo.

Imaging

Slice selection is achieved by applying a magnetic gradient in addition to the external magnetic field during the radio frequency pulse. Only one plane within the object will have protons that are on–resonance and contribute to the signal.

A real image can be considered as being composed of a number of spatial frequencies at different orientations. A two–dimensional Fourier transformation of a real image will express these waves as a matrix of spatial frequencies known as k–space.

Low spatial frequencies are represented at the centre of k–space and high spatial frequencies at the periphery. Frequency and phase encoding are used to measure the amplitudes of a range of spatial frequencies within the object being imaged.

The frequency encoding gradient is applied during readout of the signal and is orthogonal to the slice selection gradient. During application of the gradient the frequency differences in the readout direction progressively change. At the midpoint of the readout these differences are small and the low spatial frequencies in the image are sampled filling the centre of k-space.

Higher spatial frequencies will be sampled towards the beginning and end of the readout filling the periphery of k-space. Phase encoding is applied in the remaining orthogonal plane and uses the same principle of sampling the object for different spatial frequencies. However, it is applied for a brief period before the readout and the strength of the gradient is changed incrementally between each radio frequency pulse. For each phase encoding step a line of k–space is filled.

Magnet

The magnet is the largest and most expensive component of the scanner, and the remainder of the scanner is built around it. Just as important as the strength of the main magnet is its precision. The straightness of magnet lines within the centre or, as it is known as, the iso-centre of the magnet, needs to be nearly perfect.

This is known as homogeneity. Fluctuations (non-homogeneities in the field strength) within the scan region should be less than three parts per million (3 ppm). Three types of magnet have been used:

- Permanent magnet: Conventional magnets made from ferromagnetic materials (e.g., steel) can be used to provide the static magnetic field. These are extremely bulky (the magnet can weigh in excess of 100 tonnes), but once installed require little costly maintenance. Permanent magnets can only achieve limited field strength (usually < 0.4 T) and have limited stability and precision. There are also potential safety issues, as the magnetic field cannot be removed in case of entrapment.
- Resistive electromagnet: A solenoid wound from copper wire is an alternative to a permanent magnet. The advantages are low cost, but field strength is limited, and stability is poor. The electromagnet requires considerable electrical energy during operation which can make it expensive to operate. This design is essentially obsolete.
- Superconducting electromagnet: When a niobium-titanium alloy is cooled by liquid helium at 4 K ("269 °C, "452 °F) it becomes superconducting where it loses all resistance to flow of electrical current. By

building an electromagnet from superconducting wire, it is possible to develop extremely high field strengths, with very high stability. The construction of such magnets is extremely costly, and the cryogenic helium is expensive and difficult to handle. However, despite its cost, helium cooled superconducting magnets are the most common type found in MRI scanners today.

Most superconducting magnets have their coils of superconductive wire immersed in liquid helium, inside a vessel called a cryostat. Despite thermal insulation, ambient heat causes the helium to slowly boil off. Such magnets, therefore, require regular topping-up with helium.

Generally a cryocooler, also known as a coldhead is used to recondense some helium vapour back into the liquid helium bath. Several manufacturers now offer 'cryogenless' scanners, where instead of being immersed in liquid helium the magnet wire is cooled directly by a cryocooler.

Magnets are available in a variety of shapes. However, permanent magnets are most frequently 'C' shaped, and superconducting magnets most frequently cylindrical. However, C-shaped superconducting magnets and box-shaped permanent magnets have also been used.

Magnetic field strength is an important factor determining image quality. Higher magnetic fields increase signal-to-noise ratio, permitting higher resolution or faster scanning. However, higher field strengths require more costly magnets with higher maintenance costs, and have increased safety concerns. 1.0 - 1.5 T field strengths are a good compromise between cost and performance for general medical use. However, for certain specialist uses (e.g., brain imaging), field strengths up to 3.0 T may be desirable.

RADIO FREQUENCY SYSTEM

The radio frequency (RF) transmission system consists of a RF synthesizer, power amplifier and transmitting coil. This is usually built into the body of the scanner. The power of the transmitter is variable, but high-end scanners may have a peak output power of up to 35 kW, and be capable of sustaining average power of 1 kW. The receiver consists of the coil, pre-amplifier and signal processing system.

While it is possible to scan using the integrated coil for transmitting and receiving, if a small region is being imaged then better image quality is obtained by using a close-fitting smaller coil. A variety of coils are available which fit around parts of the body, e.g., the head, knee, wrist, or internally, e.g., the rectum.

A recent development in MRI technology has been the development of sophisticated multi-element phased array coils which are capable of acquiring multiple channels of data in parallel. This 'parallel imaging' technique uses unique acquisition schemes that allow for accelerated imaging, by replacing some of the spatial coding originating from the magnetic gradients with the spatial sensitivity of the different coil elements.

However the increased acceleration also reduces the signal-to-noise ratio and can create residual artifacts in the image reconstruction. Two frequently used parallel acquisition and reconstruction schemes are sense and grappa.

Gradients

Magnetic gradients are generated by three orthogonal coils, oriented in the x, y and z directions of the scanner. These are usually resistive electromagnets powered by sophisticated amplifiers which permit rapid and precise adjustments to their field strength and direction.

Typical gradient systems are capable of producing gradients from 20 mT/m to 100 mT/m (i.e. in a 1.5 T magnet, when a maximal z-axis gradient is applied the field strength may be 1.45 T at one end of a 1 m long bore, and 1.55 T at the other). It is the magnetic gradients that determine the plane of imaging - because the orthogonal gradients can be combined freely, any plane can be selected for imaging.

Scan speed is dependent on performance of the gradient system. Stronger gradients allow for faster imaging, or for higher resolution; similarly, gradients systems capable of faster switching can also permit faster scanning. However, gradient performance is limited by safety concerns over nerve stimulation.

In order to understand MRI contrast, it is important to have some understanding of the time constants involved in relaxation processes that establish equilibrium following RF excitation. As the high-energy nuclei relax and realign they emit energy at rates which are recorded to provide information about the material they are in.

The realignment of nuclear spins with the magnetic field is termed *longitudinal relaxation* and the time required for a certain percentage of the tissue's nuclei to realign is termed "Time 1" or T1, which is typically about 1 second at 1.5 Tesla main field strength.

T2-weighted imaging relies upon local dephasing of spins following the application of the transverse energy pulse; the *transverse* relaxation time is termed "Time 2" or T2, typically < 100 ms for tissue at 1.5 Tesla main field strength. A subtle but important variant of the T2 technique is called T2* imaging.

T2 imaging employs a spin echo technique, in which spins are refocused to compensate for local magnetic field inhomogeneities. T2* imaging is performed without refocusing. This sacrifices some image integrity (resolution) but provides additional sensitivity to relaxation processes that cause incoherence of transverse magnetization.

Applications of T2* imaging include functional MRI (fMRI) or evaluation of baseline vascular perfusion (e.g. cerebral blood flow (CBF)) and cerebral blood volume (CBV) using injected agents; in these cases, there is an inherent trade-off between image quality and detection sensitivity.

Because T2*-weighted sequences are sensitive to magnetic inhomogeneity (as can be caused by deposition of iron-containing blood-degradation products), such sequences are utilized to detect subtle areas of recent or chronic intra cranial hemorrhage ("Heme sequence").

Image contrast is created by using a selection of image acquisition parameters that weights signal by T1, T2 or T2, or no relaxation time ("proton-density images"). In the brain, T1-

weighting causes the nerve connections of white matter to appear white, and the congregations of neurons of gray matter to appear gray, while cerebrospinal fluid appears dark.

The contrast of "white matter," "gray matter'" and "cerebrospinal fluid" is reversed using T2 or T2* imaging, whereas proton-weighted imaging provides little contrast in normal subjects. Additionally, functional information (CBF, CBV, blood oxygenation) can be encoded within T1, T2, or T2.

Diffusion weighted imaging (DWI) uses very fast scans with an additional series of gradients (diffusion gradients) rapidly turned on and off. Protons from water diffusing randomly within the brain, via Brownian motion, lose phase coherence and, thus signal during application of diffusion gradients.

In a brain with an acute infarction water diffusion is impaired, and signal loss on DWI sequences is less than in normal brain. DWI is the most sensitive method of detecting cerebral infarction (stroke) and works within 30 minutes of the ictus.

Contrast Enhancement

Both T1-weighted and T2-weighted images are acquired for most medical examinations; However they do not always adequately show the anatomy or pathology. The first option is to use a more sophisticated image acquisition technique such as fat suppression or chemical-shift imaging. The other is to administer a contrast agent to delineate areas of interest.

A contrast agent may be as simple as water, taken orally, for imaging the stomach and small bowel although substances with specific magnetic properties may be used. Most commonly, a paramagnetic contrast agent (usually a gadolinium compound) is given.

Gadolinium-enhanced tissues and fluids appear extremely bright on T1-weighted images. This provides high sensitivity for detection of vascular tissues (e.g. tumors) and permits assessment of brain perfusion (e.g. in stroke). There have been concerns raised recently regarding the toxicity of gadolinium-based contrast agents and their impact on persons with impaired kidney function. Special actions may be taken, such as hemodialysis following a contrast MRI scan for renally-impaired patients.

More recently, superparamagnetic contrast agents (e.g. iron oxide nanoparticles) have become available. These agents appear very dark on T2-weighted images and may be used for liver imaging - normal liver tissue retains the agent, but abnormal areas (e.g. scars, tumors) do not.

They can also be taken orally, to improve visualization of the gastrointestinal tract, and to prevent water in the gastrointestinal tract from obscuring other organs (e.g. pancreas). Diamagnetic agents such as barium sulfate have been studied for potential use in the gastrointestinal tract, but are less frequently used.

MRI VS CT

A computed tomography (CT) scanner uses X-rays, a type of ionizing radiation, to acquire its images, making it a good tool for examining tissue composed of elements of a relatively higher atomic number than the tissue surrounding them, such as bone and calcifications (calcium based) within the body (carbon based flesh), or of structures (vessels, bowel).

MRI, on the other hand, uses non-ionizing radio frequency (RF) signals to acquire its images and is best suited for non-calcified tissue. CT may be enhanced by use of contrast agents containing elements of a higher atomic number than the surrounding flesh such as iodine or barium. Contrast agents for MRI are those which have paramagnetic properties. One example is gadolinium.

Both CT and MRI scanners can generate multiple two-dimensional cross-sections (slices) of tissue and three-dimensional reconstructions. Unlike CT, which uses only X-ray attenuation to generate image contrast, MRI has a long list of properties that may be used to generate image contrast. By variation of scanning parameters, tissue contrast can be altered and enhanced in various ways to detect different features.

MRI can generate cross-sectional images in any plane (including oblique planes). CT was limited to acquiring images in the axial (or near axial) plane in the past. The scans used to be called Computed *Axial* Tomography scans (CAT scans). However, the development of multi-detector CT scanners with near-isotropic resolution, allows the CT scanner to produce data that can be retrospectively reconstructed in any plane with minimal loss of image quality.

For purposes of tumor detection and identification, MRI is generally superior. However, CT usually is more widely available, faster, much less expensive, and may be less likely to require the person to be sedated or anesthetized. MRI is also best suited for cases when a patient is to undergo the exam several times successively in the short term, because, unlike CT, it does not expose the patient to the hazards of ionizing radiation.

Economics of MRI

MRI equipment is expensive. New 1.5 Tesla scanners often cost between $1,000,000 USD and $1,500,000 USD. New 3.0 Tesla scanners often cost between $2,000,000 and $2,300,000 USD. Construction of MRI suites can cost $500,000 USD.

For over a dozen years, MRI scanners have been significant sources of revenue for healthcare providers in the US. This is because of favorable reimbursement rates from insurers, both private and federal government programs. Insurance reimbursement has historically been provided in two components, technical for the actual performance of the MRI scan and professional for the radiologist's review of the images and/or data.

In the US, the 2007 Deficit Reduction Act (DRA) significantly reduced reimbursement rates paid by federal insurance programs for the technical component of many scans, shifting the economic landscape. Many private insurers have followed suit.

Currently, in the US, there is increasing interest in reducing the costs associated with MRI services and simultaneously improving the ability to effectively and efficiently provide MRI examination services to larger numbers of patients with the same equipment.

The k-space Tormalism

In 1983 Ljunggren and Tweig independently introduced the k-space formalism, a technique that proved invaluable in unifying different MR imaging techniques. They showed that the demodulated MR signal $S(t)$ generated by freely precessing nuclear spins in the presence of a linear magnetic field gradient G equals the Fourier transform of the effective spin density i.e.

$$S(t) = \tilde{\rho}\text{effective}\left(\vec{k}(t)\right) \equiv \int d^3x\rho(\vec{x}) \cdot e^{2\pi 2\vec{k}(t)\cdot\vec{x}}$$

where:

$$\vec{k}(t) \equiv \int_0^t \vec{G}(t')dt'$$

In other words, as time progresses the signal traces out a trajectory in k-space with the velocity vector of the trajectory proportional to the vector of the applied magnetic field gradient. By the term *effective spin density* we mean the true spin density $\rho(\vec{x})$ corrected for the effects of T_1 preparation, T_2 decay, dephasing due to field inhomogeneity, flow, diffusion, etc. and any other phenomena that affect that amount of transverse magnetization available to induce signal in the RF probe.

From the basic k-space formula, it follows immediately that we reconstruct an image $I(\vec{x})$ simply by taking the inverse Fourier transform of the sampled data viz.

$$I(\vec{x}) = \int d^3k S\left(\vec{k}(t)\right) \cdot e^{-2\pi\vec{k}(t)\cdot\vec{x}}$$

Using the k-space formalism, a number of seemingly complex ideas become simple. For example, it becomes very easy to understand the role of phase encoding (the so-called spin-warp method). In a standard spin echo or gradient echo scan, where the readout (or view) gradient is constant (e.g. G_x), a single line of k-space is scanned per RF excitation.

When the phase encoding gradient is zero, the line scanned is the k_x axis. When a non-zero phase-encoding pulse is added in between the RF excitation and the commencement of the readout gradient, this line moves up or down in k-space i.e. we scan the line k_y=constant.

The k-space formalism also makes it very easy to compare different scanning techniques. In single-shot EPI, all of k-space is scanned in a single shot, following either a sinusoidal or zig-zag trajectory. Since alternating lines of k-space are scanned in opposite directions, this must be taken into account in the reconstruction. Multi-shot EPI and fast spin echo techniques acquire only part of k-space per excitation.

In each shot, a different interleaved segment is acquired, and the shots are repeated until k-space is sufficiently well-covered. Since the data at the centre of k-space represent lower spatial frequencies than the data at the edges of k-space, the T_E value for the centre of k-space determines the image's T_2 contrast.

The importance of the centre of k-space in determining image contrast can be exploited in more advanced imaging techniques. One such technique is spiral acquisition - a rotating magnetic field gradient is applied, causing the trajectory in k-space to trace out spiral out from the centre to the edge.

Due to T_2 and T_2 * decay the signal is greatest at the start of the acquisition, hence acquiring the centre of k-space first improves contrast to noise ratio (CNR) when compared to conventional zig-zag acquisitions, especially in the presence of rapid movement.

Since and are conjugate variables (with respect to the Fourier transform) we can use the Nyquist theorem to show that the step in k-space determines the field of view of the image (maximum frequency that is correctly sampled) and the maximum value of k sampled determines the resolution i.e.

$$\text{For} \propto \frac{1}{\Delta k} \quad \text{Resolution} \propto |k_{max}|$$

(these relationships apply to each axis [X, Y, and Z] independently).

Chapter 16

Nuclear Power Reactors

A nuclear reactor produces and controls the release of energy from splitting the atoms of certain elements. In a nuclear power reactor, the energy released is used as heat to make steam to generate electricity. (In a research reactor the main purpose is to utilise the actual neutrons produced in the core. In most naval reactors, steam drives a turbine directly for propulsion.)

The principles for using nuclear power to produce electricity are the same for most types of reactor. The energy released from continuous fission of the atoms of the fuel is harnessed as heat in either a gas or water, and is used to produce steam. The steam is used to drive the turbines which produce electricity (as in most fossil fuel plants). There are several components common to most types of reactors:

Fuel. Usually pellets of uranium oxide (UO_2) arranged in tubes to form fuel rods. The rods are arranged into fuel assemblies in the reactor core. In a new reactor with new fuel a neutron source is needed to get the reaction going. Usually this is beryllium mixed with polonium, radium or other alpha-emitter. Alpha particles from the decay cause a release of neutrons from the beryllium as it turns to carbon-12. Restarting a reactor with some used fuel may not require this, as there may be enough neutrons to achieve criticality when control rods are removed.

Moderator. This is material which slows down the neutrons released from fission so that they cause more fission. It is usually water, but may be heavy water or graphite. Control rods. These are made with neutron-absorbing material such as cadmium, hafnium or boron, and are inserted or withdrawn from the core to control the rate of reaction, or to halt it. (Secondary shutdown systems involve adding other neutron absorbers, usually as a fluid, to the system.)

In fission, most of the neutrons are released promptly, but some are delayed. These are crucial in enabling a chain reacting system (or reactor) to be controllable and to be able to be held precisely critical.

Nuclear power plants in commercial operation

Reactor type	Main Countries	Number	GWe	Fuel	Coolant	Moderator
Pressurised Water Reactor (PWR)	US, France, Japan, Russia	264	250.5	enriched UO_2	water	water
Boiling Water Reactor (BWR)	US, Japan, Sweden	94	86.4	enriched UO_2	water	water
Pressurised Heavy Water Reactor 'CANDU' (PHWR)	Canada	43	23.6	natural UO_2	heavy water	heavy water
Gas-cooled Reactor (AGR & Magnox)	UK	18	10.8	natural U (metal), enriched UO_2	CO_2	graphite
Light Water Graphite Reactor (RBMK)	Russia	12	12.3	enriched UO_2	water	graphite
Fast Neutron Reactor (FBR)	Japan, France, Russia	4	1.0	PuO_2 and UO_2	liquid sodium	none
Other	Russia	4	0.05	enriched UO_2	water	graphite
	TOTAL	439	384.6			

Coolant. A liquid or gas circulating through the core so as to transfer the heat from it.. In light water reactors the water moderator functions also as primary coolant.

Except in BWRs, there is secondary coolant circuit where the steam is made. Pressure vessel or pressure tubes. Usually a robust steel vessel containing the reactor core and moderator/ coolant, but it may be a series of tubes holding the fuel and conveying the coolant through the moderator. Steam generator. Part of the cooling system where the heat from the reactor is used to make steam for the turbine.

Containment. The structure around the reactor core which is designed to protect it from outside intrusion and to protect those outside from the effects of radiation in case of any malfunction inside. It is typically a metre-thick concrete and steel structure. There are several different types of reactors as indicated in the following table.

GWe = capacity in thousands of megawatts (gross)

Reactor type	Main Countries	Number	GWe	Fuel	Coolant	Moderator
Pressurised Water Reactor (PWR)	US, France, Japan, Russia	264	250.5	enriched UO_2	water	water
Boiling Water Reactor (BWR)	US, Japan, Sweden	94	86.4	enriched UO_2	water	water
Pressurised Heavy Water Reactor 'CANDU' (PHWR)	Canada	43	23.6	natural UO_2	heavy water	heavy water
Gas-cooled Reactor (AGR & Magnox)	UK	18	10.8	natural U (metal), enriched UO_2	CO_2	graphite
Light Water Graphite Reactor (RBMK)	Russia	12	12.3	enriched UO_2	water	graphite
Fast Neutron Reactor (FBR)	Japan, France, Russia	4	1.0	PuO_2 and UO_2	liquid sodium	none
Other	Russia	4	0.05	enriched UO_2	water	graphite
	TOTAL	439	384.6			

GWe = capacity in thousands of megawatts (gross)

Most reactors need to be shut down for refuelling, so that the pressure vessel can be opened up. In this case refuelling is at intervals of 1-2 years, when a quarter to a third of the fuel assemblies are replaced with fresh ones. The CANDU and RBMK types have pressure tubes (rather than a pressure vessel enclosing the reactor core) and can be refuelled under load by disconnecting individual pressure tubes.

If graphite or heavy water is used as moderator, it is possible to run a power reactor on natural instead of enriched uranium. Natural uranium has the same elemental composition as when it was mined (0.7 per cent U-235, over 99.2 per cent U-238), enriched uranium has had the proportion of the fissile isotope (U-235) increased by a process called enrichment, commonly to 3.5 - 5.0 per cent.

In this case the moderator can be ordinary water, and such reactors are collectively called light water reactors. Because the light water absorbs neutrons as well as slowing them, it is less efficient as a moderator than heavy water or graphite. Practically all fuel is ceramic uranium oxide (UO_2 with a melting point of 2800°C) and most is enriched. The fuel pellets (usually about 1 cm diameter and 1.5 cm long) are typically arranged in a long zirconium alloy (zircaloy) tube to form a fuel rod, the zirconium being hard, corrosion-resistant and permeable to neutrons. Numerous rods form a fuel assembly, which is an open lattice and can be lifted into and out of the reactor core. In the most common reactors these are about 3.5 to 4 metres long.

Zirconium is an important mineral for nuclear power, where it finds its main use. It is therefore subject to controls on trading. It is normally contaminated with hafnium, a neutron absorber, so very pure 'nuclear grade' Zr is used to make the zircaloy, which is about 98 per cent Zr plus tin, iron, chromium and sometimes nickel to enhance its strength.

Burnable poisons are often used (especially in BWR) in fuel or coolant to even out the performance of the reactor over time from fresh fuel being loaded to refuelling. These are neutron absorbers which decay under neutron exposure, compensating for the progressive build up of neutron absorbers in the fuel as it is burned.

The best known is gadolinium, which is a vital ingredient of fuel in naval reactors where installing fresh fuel is very inconvenient, so reactors are designed to run more than a decade between refuellings.

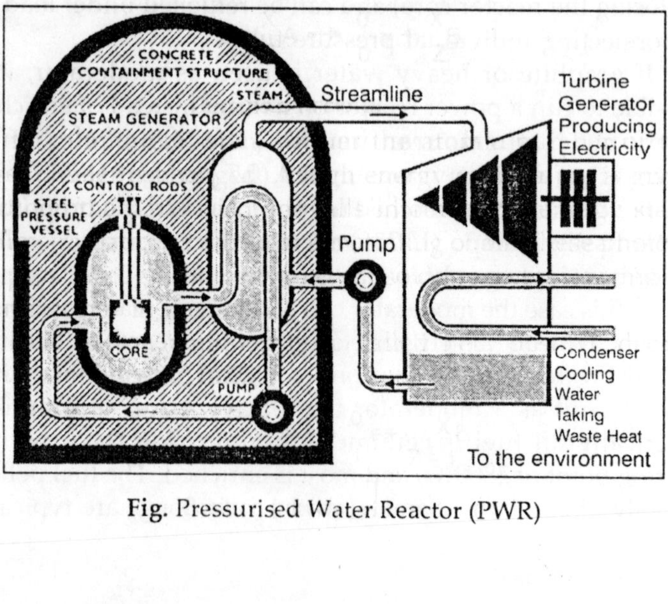

Fig. Pressurised Water Reactor (PWR)

This is the most common type, with over 230 in use for power generation and a further several hundred in naval propulsion. The design originated as a submarine power plant. It uses ordinary water as both coolant and moderator. The design is distinguished by having a primary cooling circuit which flows through the core of the reactor under very high pressure, and a secondary circuit in which steam is generated to drive the turbine.

A PWR has fuel assemblies of 200-300 rods each, arranged vertically in the core, and a large reactor would have about 150-250 fuel assemblies with 80-100 tonnes of uranium.

Water in the reactor core reaches about 325°C, hence it must be kept under about 150 times atmospheric pressure to prevent it boiling. Pressure is maintained by steam in a pressuriser.

In the primary cooling circuit the water is also the moderator, and if any of it turned to steam the fission reaction would slow down. This negative feedback effect is one of the safety features of the type. The secondary shutdown system involves adding boron to the primary circuit.

The secondary circuit is under less pressure and the water here boils in the heat exchangers which are thus steam generators. The steam drives the turbine to produce electricity, and is then condensed and returned to the heat exchangers in contact with the primary circuit.

BOILING WATER REACTOR (BWR)

This design has many similarities to the PWR, except that there is only a single circuit in which the water is at lower pressure (about 75 times atmospheric pressure) so that it boils in the core at about 285°C. The reactor is designed to operate with 12-15 per cent of the water in the top part of the core as steam, and hence with less moderating effect and thus efficiency there.

The steam passes through drier plates (steam separators) above the core and then directly to the turbines, which are thus part of the reactor circuit. Since the water around the core of a reactor is always contaminated with traces of radionuclides, it means that the turbine must be shielded and radiological protection provided during maintenance. The cost of this tends to balance the savings due to the simpler design. Most of the radioactivity in the water is very short-lived (mostly N-16, with a 7 second half-life), so the turbine hall can be entered soon after the reactor is shut down.

A BWR fuel assembly comprises 90-100 fuel rods, and there are up to 750 assemblies in a reactor core, holding up to 140 tonnes of uranium. The secondary control system involves restricting water flow through the core so that steam in the top part means moderation is reduced.

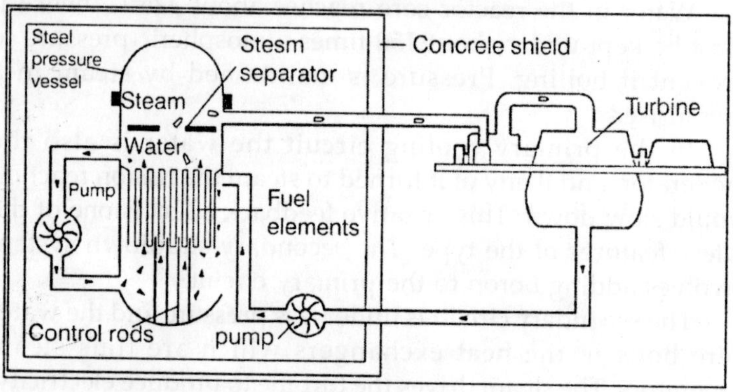

Fig. Pressurised Heavy Water Reactor (PHWR or CANDU)

The CANDU reactor design has been developed since the 1950s in Canada. It uses natural uranium (0.7 per cent U-235) oxide as fuel, hence needs a more efficient moderator, in this case heavy water (D_2O).

The moderator is in a large tank called a calandria, penetrated by several hundred horizontal pressure tubes which form channels for the fuel, cooled by a flow of heavy water under high pressure in the primary cooling circuit, reaching 290°C.

As in the PWR, the primary coolant generates steam in a secondary circuit to drive the turbines. The pressure tube design means that the reactor can be refuelled progressively without shutting down, by isolating individual pressure tubes from the cooling circuit.

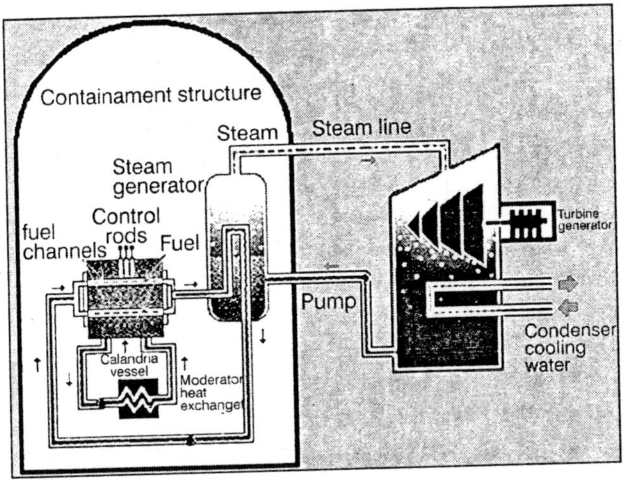

Fig. A CANDU Fuel Assembly

A CANDU fuel assembly consists of a bundle of 37 half metre long fuel rods (ceramic fuel pellets in zircaloy tubes) plus a support structure, with 12 bundles lying end to end in a fuel channel.

Control rods penetrate the calandria vertically, and a secondary shutdown system involves adding gadolinium to the moderator. The heavy water moderator circulating through the body of the calandria vessel also yields some heat.

ADVANCED GAS-COOLED REACTOR (AGR)

These are the second generation of British gas-cooled reactors, using graphite moderator and carbon dioxide as coolant. The fuel is uranium oxide pellets, enriched to 2.5-3.5 per cent, in stainless steel tubes.

The carbon dioxide circulates through the core, reaching 650°C and then past steam generator tubes outside it, but still inside the concrete and steel pressure vessel. Control rods penetrate the moderator and a secondary shutdown system involves injecting nitrogen to the coolant.

The AGR was developed from the Magnox reactor, also graphite moderated and CO_2 cooled, and a number of these are still operating in UK. They use natural uranium fuel in metal form.

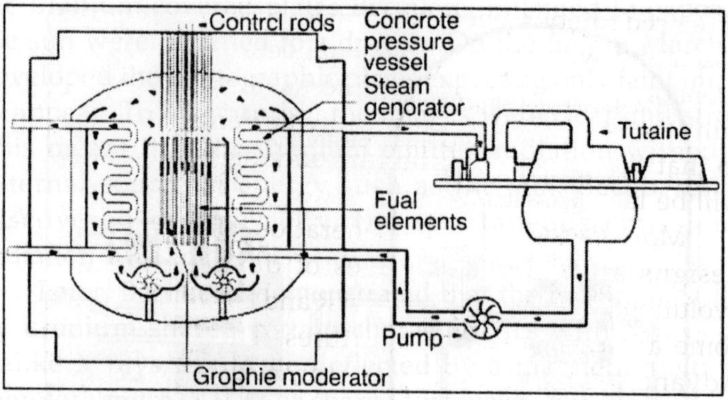

Fig. Advanced Gas-cooled Reactor

LIGHT WATER GRAPHITE-MODERATED REACTOR (RBMK)

This is a Soviet design, developed from plutonium production reactors. It employs long (7 metre) vertical pressure tubes running through graphite moderator, and is cooled by water, which is allowed to boil in the core at 290°C, much as in a BWR.

Fuel is low-enriched uranium oxide made up into fuel assemblies 3.5 metres long. With moderation largely due to the fixed graphite, excess boiling simply reduces the cooling and neutron absorbtion without inhibiting the fission reaction, and a positive feedback problem can arise.

ADVANCED REACTORS

Several generations of reactors are commonly distinguished. Generation I reactors were developed in 1950-60s and very few are still running today. They mostly used natural uranium fuel and used graphite as moderator. Generation II reactors are typified by the present US fleet and most in operation elsewhere.

They typically use enriched uranium fuel and are mostly cooled and moderated by water. Generation III are the Advanced Reactors, the first few of which are in operation in Japan and others are under construction and ready to be ordered. They are developments of the second generation with enhanced safety. Generation IV designs are still on the drawing board and will not be operational before 2020 at the earliest, probably later. They will tend to have closed fuel cycles and burn the long-lived actinides now forming part of spent fuel, so that fission products are the only high-level waste. Many will be fast neutron reactors.

More than a dozen (Generation III) advanced reactor designs are in various stages of development. Some are evolutionary from the PWR, BWR and CANDU designs above, some are more radical departures. The former include the Advanced Boiling Water Reactor, a few of which are now operating with others under construction.

The best-known radical new design is the Pebble Bed Modular Reactor, using helium as coolant, at very high temperature, to drive a turbine directly. Considering the closed fuel cycle, Generation 1-3 reactors recycle plutonium (and possibly uranium), while Generation IV are expected to have full actinide recycle.

FAST NEUTRON REACTORS

Some reactors (only one in commercial service) do not have a moderator and utilise fast neutrons, generating power from plutonium while making more of it from the U-238 isotope in or around the fuel. While they get more than 60 times as much energy from the original uranium compared with the normal reactors, they are expensive to build and await resource scarcity to come into their own.

LIFETIME OF NUCLEAR REACTORS.

Most of today's nuclear plants which were originally designed for 30 or 40-year operating lives. However, with major investments in systems, structures and components lives can be extended, and in several countries there are active programs to extend operating lives. In the USA most of the more than one hundred reactors are expected to be granted licence extensions from 40 to 60 years.

This justifies significant capital expenditure in upgrading systems and components, including building in extra performance margins. Some components simply wear out, corrode or degrade to a low level of efficiency. These need to be replaced. Steam generators are the most prominent and expensive of these, and many have been replaced after about 30 years where the reactor otherwise has the prospect of running for 60 years. This is essentially an economic decision. Lesser components are more straightforward to replace as they age. In Candu reactors, pressure tube replacement has been undertaken on some plants after about 30 years operation.

A second issue is that of obsolescence. For instance, older reactors have analogue instrument and control systems. Thirdly, the properties of materials may degrade with age, particularly with heat and neutron irradiation. In respect to all these aspects, investment is needed to maintain reliability and safety. Also, periodic safety reviews are undertaken on older plants in line with international safety conventions and principles to ensure that safety margins are maintained.

FLOATING NUCLEAR POWER PLANTS

Apart from over 200 nuclear reactors powering various kinds of ships, Rosatom in Russia has set up a subsidiary to supply floating nuclear power plants ranging in size from 70 to 600 MWe. These will be mounted in pairson a large barge, which will be permanently moored where it is needed to supply power and possibly some desalination to a shore settlement or industrial complex.

The first will have two 40 MWe reactors based on those in icebreakers and will operate at Severodvinsk, in the Archangel region. Five of the next seven will be used by Gazprom for offshore oil and gas field development and for operations on the Kola and Yamal peninsulas. One is for Pevek on the Chukotka peninsula, another for Kamchatka region, both in the far east of the country.

Further far east sites being considered are Yakutia and Taimyr. Electricity cost is expected to be much lower than from present alternatives. The Russian KLT-40S is a reactor well proven in icebreakers and now proposed for wider use in desalination and, on barges, for remote area power supply.

Here a 150 MWt unit produces 35 MWe (gross) as well as up to 35 MW of heat for desalination or district heating. These are designed to run 3-4 years between refuelling and it is envisaged that they will be operated in pairs to allow for outages, with on-board refuelling capability and used fuel storage.

At the end of a 12-year operating cycle the whole plant is taken to a central facility for overhaul and removal of used fuel. Two units will be mounted on a 20,000 tonne barge. A larger Russian factory-built and barge-mounted reactor is the VBER-150, of 350 MW thermal, 110 MWe. The larger VBER-300 PWR is a 325 MWe unit, originally envisaged in pairs as a floating nuclear power plant, displacing 49,000 tonnes. As a cogeneration plant it is rated at 200 MWe and 1900 GJ/hr.

PRIMARY COOLANTS

The advent of some of the designs mentioned above provides opportunity to review the various primary coolants used in nuclear reactors:

Water or heavy water must be maintained at very high pressure (1000-2200 psi, 7-15 MPa) to enable it to function above 100°C, as in present reactors. This has a major influence on reactor engineering. However, supercritical water around 25 MPa can give 45 per cent thermal efficiency - as at some fossil-fuel power plants today with outlet temperatures of 600°C, and at ultra supercritical levels (30+ MPa) 50 per cent may be attained.

Helium must be used at similar pressure (1000-2000 psi, 7-14 MPa) to maintain sufficient density for efficient operation. Again, there are engineering implications, but it can be used in the Brayton cycle to drive a turbine directly. Carbon dioxide was used in early British reactors and their AGRs. It is denser than helium and thus likely to give better thermal conversion efficiency.

There is now interest in supercritical CO_2 for the Brayton cycle. Sodium, as normally used in fast neutron reactors, melts at 98°C and boils at 883°C at atmospheric pressure, so despite the need to keep it dry the engineering required to contain it is relatively modest. However, normally water/steam is used in the secondary circuit to drive a turbine (Rankine cycle) at lower thermal efficiency than the Brayton cycle. Lead or lead-bismuth are capable of higher temperature operation. They are transparent to neutrons, aiding efficiency, and do not react with water.

However, they are corrosive of fuel cladding and steels, and Pb-Bi yields Po activation products. Pb-Bi melts at 125°C and boils at 1670°C, Pb melts at 327°C and boils at 1737°C. In 1998 Russia declassified a lot of research information derived from its experience with submarine reactors, and US interest in using Pb/Pb-Bi for small reactors has increased subsequently. Molten fluoride salt boils at 1400°C at atmospheric pressure, so allows several options for use of the heat, including using helium in a secondary Brayton cycle with thermal efficiencies of 48 per cent at 750°C to 59 per cent at 1000°C, or manufacture of hydrogen.

Low-pressure liquid coolants allow all their heat to be delivered at high temperatures, since the temperature drop in heat exchangers is less than with gas coolants. Also, with a good margin between operating and boiling temperatures, passive cooling for decay heat is readily achieved. The removal of passive decay heat is a vital feature of primary cooling systems, beyond heat transfer to do work.

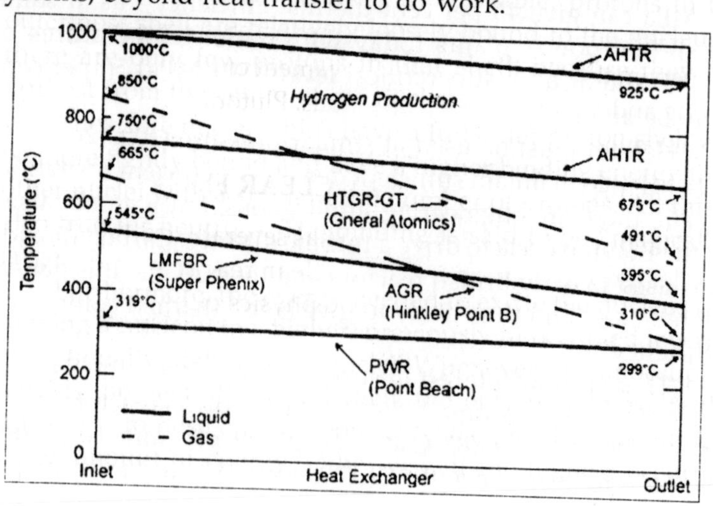

When the fission process stops, fission product decay continues and a substantial amount of heat is added to the core. At the moment of shutdown, this is about 6 per cent of the full power level, but it quickly drops to about 1 per cent as the short-lived fission products decay. This heat could melt the core of a light water reactor unless it is reliably dissipated. Typically some kind of convection flow is relied upon.

PRIMITIVE REACTORS

The world's oldest known nuclear reactors operated at what is now Oklo in Gabon, West Africa. About 2 billion years ago, at least 17 natural nuclear reactors achieved criticality in a rich deposit of uranium ore. Each operated at about 20 kW thermal. At that time the concentration of U-235 in all natural uranium was 3.7 per cent instead of 0.7 per cent as at present. (U-235 decays much faster than U-238, whose half-life is about the same as the age of the Earth.) These natural chain reactions, started spontaneously by the presence of water acting as a moderator, continued for about 2 million years before finally dying away.

During this long reaction period about 5.4 tonnes of fission products as well as 1.5 tonnes of plutonium together with other transuranic elements were generated in the orebody. The initial radioactive products have long since decayed into stable elements but close study of the amount and location of these has shown that there was little movement of radioactive wastes during and after the nuclear reactions. Plutonium and the other transuranics remained immobile.

THE PHYSICS OF NUCLEAR FISSION

The physics of nuclear fission has several quirks that affect the design and behaviour of nuclear reactors; this article presents a general overview of the physics of nuclear reactors and their behaviour.

CRITICALITY

In a nuclear reactor, the neutron population at any instant is a function of the rate of neutron production (due to fission processes) and the rate of neutron losses (via non-fission absorption mechanisms and leakage from the system). When a reactor's neutron population remains steady from one generation to the next, the fission chain reaction is self-sustaining and the reactor's condition is referred to as "CRITICAL".

When the reactor's neutron production exceeds losses, characterized by increasing power level, it's called "SUPERcritical", and; when losses dominate it's "SUBcritical" & exhibits decreasing power.

The "six-factor formula" is the neutron life-cycle balance equation, which includes six separate factors, the product of which is equal to the ratio of the number of neutrons in any generation to that of the previous one; this parameter is called the effective multiplication factor (*k*), *a.k.a.*

K_{eff}, $k = L_f \rho L_{th} f \eta c$, where L_f = "fast non-leakage factor"; ρ = "resonance escape probability"; L_{th} = "thermal non-leakage factor"; *f* = "thermal fuel utilization factor"; η = "reproduction factor"; *c* = "fast-fission factor".

k = (Neutrons produced in one generation)/(Neutrons produced in the previous generation) When the reactor is critical, $k = 1$. When the reactor is subcritical, $k < 1$. When the reactor is supercritical, $k > 1$.

"Reactivity" is an expression of the departure from criticality. äk = $(k - 1)/k$ When the reactor is critical, $\delta k = 0$. When the reactor is subcritical, $\delta k < 0$. When the reactor is supercritical, $\delta k > 0$. Reactivity is also represented by the lowercase Greek letter rho (ρ).

If we write 'N' for the number of free neutrons in a reactor core and 'τ' for the average lifetime of each neutron (before it either escapes from the core or is absorbed by a nucleus), then the reactor will follow differential equation (the *evolution equation*)

$$dN / dt = \alpha N / \tau$$

where á is a constant of proportionality, and dN / dt is the rate of change of the neutron count in the core. This type of differential equation describes exponential growth or exponential decay, depending on the sign of the constant á, which is just the expected number of neutrons after one average neutron lifetime has elapsed:

$$\alpha = P_{impact}P_{fission}n_{avg} - P_{absorb} - P_{escape}$$

Here, P_{impact} is the probability that a particular neutron will strike a fuel nucleus, $P_{fission}$ is the probability that the neutron, having struck the fuel, will cause that nucleus to undergo fission, P_{absorb} is the probability that it will be absorbed by something other than fuel, and P_{escape} is the probability that it will "escape" by leaving the core altogether. n_{avg} is the number of neutrons produced, on average, by a fission event — it is between 2 and 3 for both ^{235}U and ^{239}Pu.

If á is positive, then the core is *supercritical* and the rate of neutron production will grow exponentially until some other effect stops the growth. If á is negative, then the core is "subcritical" and the number of free neutrons in the core will shrink exponentially until it reaches an equilibrium at zero (or the background level from spontaneous fission). If á is exactly zero, then the reactor is *critical* and its output does not vary in time ($dN / dt = 0$, from above).

Nuclear reactors are engineered to reduce P_{escape} and P_{absorb}. Small, compact structures reduce the probability of direct escape by minimizing the surface area of the core, and some materials (such as graphite) can reflect some neutrons back into the core, further reducing P_{escape}. Light metals such as aluminum that are not strong neutron absorbers are used to build the structure of reactor cores.

The probability of fission, $P_{fission}$, depends on the nuclear physics of the fuel, and is often expressed as a cross section. Reactors are usually controlled by adjusting P_{absorb}. control rods made of a strongly neutron-absorbent material such as cadmium or boron can be inserted into the core: any neutron that happens to impact the control rod is lost from the chain reaction, reducing á. P_{absorb} is also controlled by the recent history of the reactor core itself.

STARTER SOURCES

The mere fact that an assembly is supercritical does not guarantee that it contains any free neutrons at all. At least one neutron is required to "strike" a chain reaction, and if the spontaneous fission rate is sufficiently low it may take a long time (in ^{235}U reactors, as long as many minutes) before a chance neutron encounter starts a chain reaction even if the reactor is supercritical. Most nuclear reactors include a "starter" neutron source that ensures there are always a few free neutrons in the reactor core, so that a chain reaction will begin immediately when the core is made critical.

A common type of neutron source is a mixture of an alpha particle emitter such as ^{241}Am (Americium-241) with a lightweight isotope such as ^{9}Be (Beryllium-9). Once the chain reaction is begun, the starter source is removed from the core to prevent damage from the high neutron flux in the operating reactor core.

SUBCRITICAL MULTIPLICATION

Even in a subcritical assembly such as a shut-down reactor core, any stray neutron that happens to be present in the core (for example from spontaneous fission of the fuel, from radioactive decay of fission products, or from a neutron source) will trigger an exponentially decaying chain reaction.

Although the chain reaction is not self-sustaining, it acts as a multiplier that increases the equilibrium number of neutrons in the core. This *subcritical multiplication* effect can be used in two ways: as a probe of how close a core is to criticality, and as a way to generate fission power without the risks associated with a critical mass.

As a measurement technique, subcritical multiplication was used during the Manhattan Project in early experiments to determine the minimum critical masses of ^{235}U and of ^{239}Pu. It is still used today to calibrate the controls for nuclear reactors during startup, as many effects can change the required control settings to achieve criticality in a reactor.

As a power-generating technique, subcritical multiplication allows generation of nuclear power for fission where a critical assembly is undesirable for safety or other reasons. A subcritical assembly together with a neutron source can serve as a steady source of heat to generate power from fission. Including the effect of an external neutron source ("external" to the fission process, not physically external to the core), one can write a modified evolution equation:

$$dN / dt = \alpha N / \tau + R_{ext}$$

where R_{ext} is the rate at which the external source injects neutrons into the core. In equilibrium, the core is not changing and dN/dt is zero, so the equilibrium number of neutrons is given by:

$$N = \tau R_{ext} / (-\alpha)$$

If the core is subcritical, then á is negative so there is an equilibrium with a positive number of neutrons. If the core is close to criticality, then á is very small and thus the final number of neutrons can be made arbitrarily large.

NEUTRON MODERATORS

To improve $P_{fission}$ and enable a chain reaction, uranium-fueled reactors must include a neutron moderator that interacts with newly produced fast neutrons from fission events to reduce their kinetic energy from several MeV to several eV, making them more likely to induce fission. This is because ^{235}U is much more likely to undergo fission when struck by one of these thermal neutrons than by a freshly-produced neutron from fission.

Neutron moderators are materials that interact weakly with the neutrons but absorb kinetic energy from them. Most moderators rely on either weakly bound hydrogen or a loose crystal structure of another light element such as carbon to transfer kinetic energy from the fast-moving neutrons.

Hydrogen moderators include water (H_2O), heavy water(D_2O), and zirconium hydride (ZrH_2), all of which work because a hydrogen nucleus has nearly the same mass as a free neutron: neutron-H_2O or neutron-ZrH_2 impacts excite rotational modes of the molecules (spinning them around). Deuterium nuclei (in heavy water) absorb kinetic energy less well than do light hydrogen nuclei, but they are much less likely to absorb the impacting neutron.

Water or heavy water have the advantage of being transparent liquids, so that, in addition to shielding and moderating a reactor core, they permit direct viewing of the core in operation and can also serve as a working fluid for heat transfer.

Crystal structure moderators rely on a floppy crystal matrix to absorb phonons from neutron-crystal impacts. Graphite is the most common example of such a moderator. It was used in Chicago Pile-1, the world's first man-made critical assembly, and was commonplace in early reactor designs including the Soviet RBMK nuclear power plants, of which the Chernobyl plant was one.

MODERATORS AND REACTOR DESIGN

The amount and nature of neutron moderation affects reactor controllability and hence safety. Because moderators both slow and absorb neutrons, there is an optimum amount of moderator to include in a given geometry of reactor core. Less moderation reduces the effectiveness by reducing the $P_{fission}$ term in the evolution equation, and more moderation reduces the effectiveness by increasing the P_{escape} term.

Most moderators become less effective with increasing temperature, so *under-moderated* reactors are stable against changes in temperature in the reactor core: if the core overheats, then the quality of the moderator is reduced and the reaction tends to slow down (there is a "negative temperature coefficient" in the reactivity of the core).

Water is an extreme case: in extreme heat, it can boil, producing effective voids in the reactor core without destroying the physical structure of the core; this tends to shut down the reaction and reduce the possibility of a fuel meltdown. *Over-moderated* reactors are unstable against changes in temperature (there is a "positive temperature coefficient" in the reactivity of the core), and so are less inherently safe than under-moderated cores.

Most reactors in use today use a combination of moderator materials. For example, TRIGA type research reactors use ZrH_2 moderator mixed with the ^{235}U fuel, an H_2O-filled core, and C (graphite) moderator and reflector blocks around the periphery of the core.

Delayed Neutrons and Controllability

Fission reactions and subsequent neutron escape happen very quickly; this is important for nuclear weapons, where the object is to make a nuclear core release as much energy as possible before it physically explodes. Most neutrons emitted by fission events are prompt: they are emitted essentially instantaneously.

Once emitted, the average neutron lifetime (τ) in a typical core is on the order of a millisecond, so if the exponential factor á is as small as 0.01, then in one second the reactor power will vary by a factor of $(1 + 0.01)^{1000}$, or more than ten thousand. Nuclear weapons are engineered to maximize the power growth rate, with lifetimes well under a millisecond and exponential factors close to 2; but such rapid variation would render it practically impossible to control the reaction rates in a nuclear reactor.

Fortunately, the *effective* neutron lifetime is much longer than the average lifetime of a single neutron in the core. About 0.65 per cent of the neutrons produced by ^{235}U fission, and about 0.75 per cent of the neutrons produced by ^{239}Pu fission, are not produced immediately, but rather are emitted by radioactive decay of fission products, with an average lifetime of about 15 seconds.

These delayed neutrons increase the effective average lifetime of neutrons in the core, to nearly 0.1 seconds, so that a core with á of 0.01 would increase in one second by only a factor of $(1 + 0.01)^{10}$, or about 1.1 — a 10 per cent increase. This is a controllable rate of change.

Most nuclear reactors are hence operated in a *prompt subcritical, delayed critical* condition: the prompt neutrons alone are not sufficient to sustain a chain reaction, but the delayed neutrons make up the small difference required to keep the reaction going.

This has effects on how reactors are controlled: when a small amount of control rod is slid into or out of the reactor core, the power level changes at first very rapidly due to *prompt subcritical multiplication* and then more gradually, following the exponential growth or decay curve of the delayed critical reaction. Further, *increases* in reactor power can be performed at any desired rate simply by pulling out a sufficient length of control rod — but *decreases* are limited in speed, because even if the reactor is taken deeply subcritical, the delayed neutrons are produced by ordinary radioactive decay of fission products and that decay cannot be hastened.

REACTOR POISONS

Any element that strongly absorbs neutrons is called a reactor poison, because it tends to shut down (poison) an ongoing fission chain reaction. Some reactor poisons are deliberately inserted into fission reactor cores to control the reaction; boron or cadmium control rods are the best example. Many reactor poisons are produced by the fission process itself, and buildup of neutron-absorbing fission products affects both the fuel economics and the controllability of nuclear reactors.

Long-lived Poisons and Fuel Reprocessing

In practice, buildup of reactor poisons in nuclear fuel is what determines the lifetime of nuclear fuel in a reactor: long before all possible fissions have taken place, buildup of long-lived neutron absorbing fission products damps out the chain reaction.

This is the reason that nuclear reprocessing is a useful activity: spent nuclear fuel contains about 99 per cent of the original fissionable material present in newly manufactured nuclear fuel. Chemical separation of the fission products restores the nuclear fuel so that it can be used again.

Nuclear reprocessing is useful economically because chemical separation is much simpler to accomplish than the difficult isotope separation required to prepare nuclear fuel from natural uranium ore, so that in principle chemical separation yields more generated energy for less effort than mining, purifying, and isotopically separating new uranium ore.

In practice, both the difficulty of handling the highly radioactive fission products and other political concerns make fuel reprocessing a contentious subject. One such concern is the fact that spent uranium nuclear fuel contains significant quantities of ^{239}Pu, a prime ingredient in nuclear weapons.

Short-lived Poisons and Controllability

Short-lived reactor poisons in fission products strongly affect how nuclear reactors can operate. Unstable fission product nuclei transmute into many different elements (*secondary fission products*) as they undergo a decay chain to a stable isotope.

The most important such element is xenon, because the isotope ^{135}Xe, a secondary fission product with a half-life of about 9 hours, is an extremely strong neutron absorber. In an operating reactor, each nucleus of ^{135}Xe is destroyed by neutron capture almost as soon as it is created, so that there is no buildup in the core.

However, when a reactor shuts down, the level of ^{135}Xe builds up in the core for about 9 hours before beginning to decay. The result is that, about 6-8 hours after a reactor is shut down, it can become physically impossible to restart the chain reaction until the ^{135}Xe has had a chance to decay over the next several hours; this is one reason why nuclear power reactors are best operated at an even power level around the clock.

^{135}Xe buildup in a reactor core makes it extremely dangerous to operate the reactor a few hours after it has been shut down. Because the ^{135}Xe absorbs neutrons strongly, starting a reactor in a high-Xe condition requires pulling the control rods out of the core much farther than normal.

But if the reactor does achieve criticality, then the neutron flux in the core will become quite high and the ^{135}Xe will be destroyed rapidly — this has the same effect as very rapidly removing a great length of control rod from the core, and can cause the reaction to grow too rapidly or even become prompt critical.

^{135}Xe played a large part in the Chernobyl accident: about eight hours after a scheduled maintenance shutdown, workers tried to bring the reactor to a zero power critical condition to test a control circuit, but because the core was loaded with ^{135}Xe from the previous day's power generation, the reaction rapidly grew uncontrollably, leading to steam explosion in the core, fire, and violent destruction of the facility.

Uranium Enrichment

While many fissionable isotopes exist in nature, the only usefully fissile isotope found in any quantity is ^{235}U; about 0.7 per cent of the uranium in most ores is the 235 isotope, and about 99.3 per cent is the inert 238 isotope. For most uses as a nuclear fuel, uranium must be *enriched* - purified so that it contains a higher percentage of 235.

Because ^{238}U absorbs fast neutrons, uranium nuclear weapons require their uranium fuel to be ~90 per cent ^{235}U to work. Nuclear reactors with water moderation can operate with only moderate enrichment of ~5 per cent ^{235}U.

Nuclear reactors with heavy water moderation can operate with natural uranium, eliminating altogether the need for enrichment and preventing the fuel from being useful for nuclear weapons; the CANDU power reactors used in Canadian power plants are an example of this type.

Uranium enrichment is extremely difficult, because the chemical properties of ^{235}U and ^{238}U are identical, so physical processes such as gas diffusion or mass spectrometry must be used to separate the isotopes based on their slightly different mass.

Because enrichment is the main technical hurdle to production of nuclear fuel and simple nuclear weapons, enrichment technology is politically sensitive.

Oklo: a Natural Nuclear Reactor

Modern deposits of uranium contain only up to ~0.7 per cent ^{235}U (and ~99.3 per cent ^{238}U), which is not enough to sustain a chain reaction moderated by ordinary water. But ^{235}U has a much shorter half-life (700 million years) than ^{238}U (4.5 billion years), so in the distant past the percentage of ^{235}U was much higher.

About two billion years ago, a water-saturated uranium deposit (in what is now the Oklo mine in Gabon, West Africa) underwent a naturally occurring chain reaction that was moderated by groundwater and, presumably, controlled by the negative void coefficient as the water boiled from the heat of the reaction.

Uranium from the Oklo mine is about 50 per cent depleted compared to other locations: it is only about 0.3 per cent to 0.7 per cent ^{235}U; and the ore contains traces of stable daughters of long-decayed fission products.

COOLING POWER PLANTS

The most common types of nuclear power plants use water for cooling in two ways: to convey heat from the reactor core to the steam turbines, and to remove and dump surplus heat from this steam circuit. (In any steam cycle plant there is a loss of about two thirds of the energy due to the intrinsic limitations of turning heat into mechanical energy.)

Combined cycle gas turbine plants need only about one third as much engineered cooling as normal thermal plants, since much heat is discharged in the turbine exhaust. The bigger the temperature difference between the internal heat source and the external environment where the surplus heat is dumped, the more efficient is the process in achieving mechanical work - here, turning a generator.

Hence the desirability of having a high temperature internally and a low temperature environmentally. In a coal-fired or conventionally gas-fired plant it is possible to run the internal boilers at higher temperatures than those with finely-engineered nuclear fuel assemblies which must avoid damage. The external consideration gives rise to desirably siting power plants alongside very cold water.

Steam Cycle Heat Transfer

For the heat transfer function the water is circulated continuously in a closed loop steam cycle and hardly any is lost.

It is turned to steam by the primary heat source to drive the turbine to do work making electricity, and it is then condensed and retuned under pressure to the heat source in a closed system. A very small amount of make-up water is required in any such system. The water needs to be clean and fairly pure. This function is much the same whether the power plant is nuclear, coal-fired, or conventionally gas-fired. Any steam cycle power plant functions in this way. At least 90 per cent of the non-hydro electricity in every country is produced thus.

Cooling to Condense the Steam and Surplus Heat Discharge

The second function for water in such a power plant is to cool the system so as to condense the low-pressure steam and recycle it. As the steam in the internal circuit condenses back to water, the surplus (waste) heat which is removed from it needs to be discharged by transfer to the air or to a body of water.

This cooling function to condense the steam is normally done in one of two ways:

- If the power plant is next to the sea, a big river or large inland water body it may be done simply by running a large amount of water through the condensers in a single pass and discharging it back into the sea, lake or river a few degrees warmer and without much loss from the amount withdrawn. That is the simplest method–"once-through" cooling. The water may be salt or fresh. Some small amount of evaporation will occur off site due to the water being a few degrees warmer.
- If the power plant does not have such abundant water, it may be done by passing it through the condenser and then using a cooling tower, where an updraught of air through water droplets cools the water. Occasionally an on-site pond may be sufficient for this. Normally the cooling is chiefly through evaporation, with simple heat transfer to the air being of less significance.

  The cooling tower evaporates up to 5 per cent of the flow and the cooled water is then returned to the condenser. The 3 to 5 per cent or so is effectively consumed, and must be continually replaced. This is the main type of recirculating cooling.

In either of these set-ups, there is no basic difference in water consumption or use between a nuclear and a coal plant. Apart from size, any differences between plants is due to thermal efficiency, ie how much heat has to be discharged into the environment, which in turn largely depends on the operating temperature in the steam generators.

A nuclear or coal plant running at 33 per cent thermal efficiency will need to dump about 14 per cent more heat than one at 36 per cent efficiency. Nuclear plants currently being built have about 34-36 per cent thermal efficiency, depending on site (especially water temperature).

Older ones are often only 32-33 per cent efficient. The relatively new Stanwell coal-fired plant in Queensland runs at 36 per cent, but some new coal-fired plants approach 40 per cent and one of the new nuclear reactors claims 39 per cent. In Europe (especially Scandinavia) low water temperature is an important criterion for power plant location.

For the planned Turkish nuclear plant, there is a one per cent gain in output if any particular plant is sited on the Black Sea coast with cooler water (average 5°C lower) than on the Mediterranean coast.

Gas combined cycle (combined cycle gas turbine - CCGT) plants need only about one third as much engineered cooling as normal thermal plants (much heat being released in the turbine exhaust), and these often use dry cooling for the second stage.

CCGT plants have an oil or gas-fired gas turbine (jet engine) coupled to a generator. The exhaust is passed through a steam generator and the steam is used to drive another turbine. This results in overall thermal efficiency of over 50 per cent.

The steam in the second phase must be condensed either with an air cooled condenser or some kind of wet cooling. Combined heat and power (CHP) plants obviously need less engineered cooling provision that others since the by-product heat is actually used for something and not dissipated uselessly.

ONCE-THROUGH WET COOLING

If a coal or nuclear plant is next to a large volume of water (big river, lake or sea), cooling can be achieved by simply running water through the plant and discharging it at a slightly higher temperature. There is then hardly any use in the sense of consumption or depletion on site, though some evaporation will occur as it cools downstream.

The amount of water required will be greater, but it is withdrawn and returned, not consumed by evaporation. Many nuclear power plants have once-through cooling, since their location is not at all determined by the source of the fuel, and depends first on where the power is needed and secondly on water availability for cooling.

Using seawater means that higher-grade materials must be used to prevent corrosion, but cooling is often more efficient. In a 2008 French government study, siting an EPR on a river instead of the coast would decrease its output by 0.9 per cent and increase the kWh cost by 3 per cent.

Any nuclear or coal-fired plant that is normally cooled by drawing water from a river or lake will have limits imposed on the temperature of the returned water (typically 30 degrees C) and/or on the temperature differential between inlet and discharge. In hot summer conditions even the inlet water from a river may approach the limit set for discharge, and this will mean that the plant is unable to run at full power.

Sometimes a supplementary cooling tower is used to help, as with many inland plants in France and Germany and at the Huntly plant in New Zealand, but this means that some water is then lost by evaporation.

RECIRCULATING WET COOLING

Where a power plant does not have abundant water, it can discharge surplus heat to the air using recirculating water systems which mostly use the physics of evaporation. Cooling towers with recirculating water are a common visual feature of power plants, often with condensed water vapour plumes. Sometimes in a cool climate it is possible to use simply a pond, from which hot water evaporates.

Most nuclear power (and other thermal) plants with recirculating cooling are cooled by water in a condenser circuit with the hot water then going to a cooling tower. This may employ either natural draft (chimney effect) or mechanical draft using large fans (enabling a lower profile). The cooling in the tower is by transferring the water's heat to the air, both directly and through evaporation of some of the water.

Concrete natural draft towers have a large concrete shell with a heat exchange 'fill' in a layer above the cold air inlet at the base of the shell. The warm air rises up through the shell by convection, creating a natural draft to provide airflow which cools the water.

Hot water is sprayed in at the top. These towers therefore do not require fans and have low operating but significant maintenance costs. They are used in large plants in Europe, eastern USA and South Africa. Mechanical draft cooling towers have large axial flow fans in a timber and plastic structure. The fans provide the airflow and are able to provide lower water temperatures than natural draft towers, particularly on hot dry days.

However, they have the disadvantage of requiring auxiliary power, typically about 0.5 per cent of the plant's output. Mechanical draft towers are used exclusively in central and western USA since they can provide a more controlled performance over a wide range of conditions, ranging from freezing to hot and dry.

Such cooling towers give rise to water consumption, with up to 2.5 litres being evaporated for each kilowatt-hour produced, depending on conditions. This sort of water consumption means that most heat is taken out of the recirculating water simply by phase change of a few per cent of it from liquid to vapour.

Cooling towers with recirculating water reportedly reduce the overall efficiency of a power plant by 3-5 per cent compared with once-through use of water from sea, lake or large stream. Due to evaporating water and thereby concentrating impurities in it, some bleed - known as "blowdown" - is needed and replacement water required is thus about 50 per cent more than actual evaporation replacement. Despite many coal and nuclear plants using wet cooling towers, in the USA electric power generation accounts for about 3 per cent of all freshwater consumption, according to the US Geological Survey - some 15.2 gigalitres per day (5550 GL/yr).

This would be simply for inland coal and nuclear plants without access to abundant water for once-through cooling. Australian coal-fired power plants consume about 400 GL/yr - the equivalent of Melbourne's water supply.

DRY COOLING

A few power plants are cooled simply by air, without relying on the physics of evaporation. These work like a automobile radiator and employ high-flow forced draft past a system of finned tubes in the condenser through which the steam passes, simply transferring its heat to the ambient air directly.

The whole power plant then uses less than 10 per cent of the water required for a wet-cooled plant, but a lot of power (around one to 1.5 per cent of power station's output) is consumed by the large fans required. This is direct dry cooling, using air-cooled condenser (ACC).

Alternatively there may still be a condenser cooling circuit as with wet recirculating cooling, but the water in it is enclosed and cooled by a flow of air past finned tubes in a cooling tower. Heat is transferred to the air, but inefficiently. This technology is not favoured if wet cooling is possible, but energy use is only 0.5 per cent of output.

In both cases there is no dependence on vaporization and hence no evaporative loss of cooling water. But the heat transfer is much less efficient and hence requires much larger cooling plant which is mechanically more complex. Eskom in South Africa quotes dry-cooled plants as having total station water consumption under 0.8 lites/kWh, this being for steam cycle losses (cf about 2.5 L/kWh for wet-cooled plants).

Both types of dry cooling involve greater cost for the cooling set-up and are much less efficient than wet cooling towers using the physics of evaporation since the only cooling is by relatively inefficient heat transfer from steam or water to air via metal fins, not by evaporation. In a hot climate the ambient air temperature may be 40°C, which severely limits the cooling potential compared with a wet bulb temperature of maybe 20°C which defines the potential for a wet system.

This loss of efficiency is reflected in greater emissions of CO_2 and pollutants per kilowatt-hour. Hybrid cooling systems combine dry and wet recirculating cooling.

NUCLEAR RELATIVE TO OTHERS, US STUDIES

It is evident that apart from any difference in thermal efficiency which affects the amount of heat to be dumped in the cooling system, there is no real difference in the amount of water used for cooling nuclear power plants, relative to coal-fired plants of the same size. However, some US studies quote a significant difference between coal and nuclear plants, this evidently being related to the (unstated) thermal efficiency of selected examples.

The studies exclude nuclear plants on the coast, which employ salt water for cooling. The second report notes that a major effect of the US Clean Water Act is to regulate the impact of cooling water use on aquatic life, and this is likely to drive the choice towards recirculating systems over once-through ones for freshwater.

This will increase water consumption unless more expensive and less efficient dry cooling systems are used. This will disadvantage nuclear over supercritical coal, though flue gas desulfurization (FGD) demands for coal will even out the water balance at least to some extent.

In France, all but four of EdF's nuclear power plants (14 reactors) are inland, and require fresh water for cooling. Eleven of the 15 inland plants (32 reactors) have cooling towers, using evaporative cooling, the other four (12 reactors) use simply river or lake water directly.

With regulatory constraints on the temperature increase in receiving waters, this means that in very hot summers generation output may be limited. In a nuclear plant, beyond some minor chlorination, the cooling water is not polluted by use - it is never in contact with the nuclear part of the plant but only cools the condenser in the turbine hall.

Future Implications for Nuclear Power

Fresh water is a valuable resource in most parts of the world. Where it is at all scarce, public opinion supports government policies, supported by common sense, to minimise the waste of it.

Apart from proximity to the main load centres, there is no reason to site nuclear power plants away from a coast, where they can use once-through seawater cooling.

Coal plant locations need to have consideration for the logistics of fuel supply (and associated aesthetics), with over three million tonnes of coal being required per year for each 1000 MWe plant.

If any thermal power plant - coal or nuclear - needs to be sited inland, the availability of cooling water is a key factor in location. Where cooling water is limited, the importance of high thermal efficiency is great, though any advantage of, say, supercritical coal over nuclear is likely to be greatly diminished due to water requirements for FGD.

Even if water is so limited that it cannot be used for cooling, there is the option of dry cooling (with extra cost) for either coal or nuclear plants. Alternatively the plant can be sited away from the load demand and where there is sufficient water for efficient cooling (accepting some losses and extra cost for transmission).

Generation III+ nuclear plants have high thermal efficiency relative to older ones, and should not be disadvantaged relative to coal by water use considerations. If water is very scarce, dry cooling can be used, though this has some cost and efficiency penalty.

Index

F

G

H

I

K

L

M